Joseph Black: a bibliography

Joseph Black

a bibliography

Compiled by J G Fyffe
and R G W Anderson

Science Museum 1992

Graeme Fyffe is Collections Development Librarian at the Science Museum Library.
Robert Anderson, fomerly Keeper of Chemistry at the Science Museum, is now Director of the British Museum.

We are grateful to Anne Jack for preparing the camera ready copy of the bibliography on an IBM Composer.

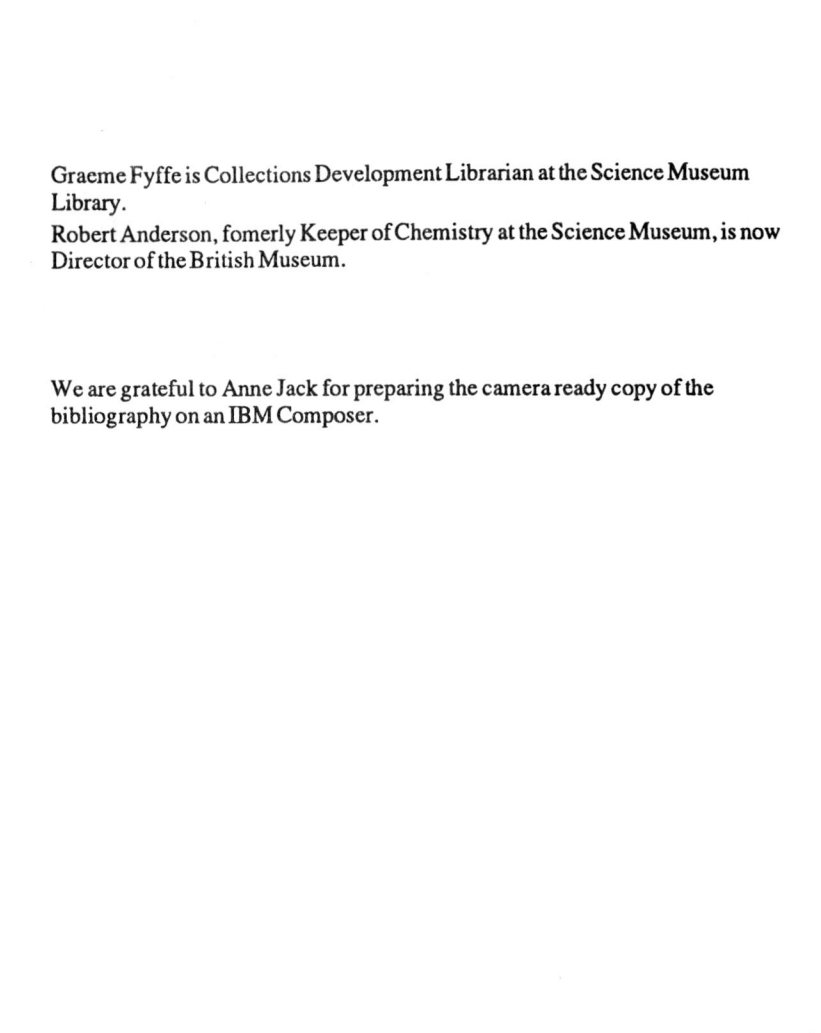

Printed in England by Antony Rowe Ltd, Chippenham, Wilts.

© 1992 Trustees of the Science Museum

All rights reserved.

ISBN 0 901805 43 2

Science Museum, Exhibition Road, London SW7 2DD

CONTENTS

                Introduction vii

Part I   Works by Black 3

Part II   Works by Black's contemporaries 29

Part III   The Black/Meyer controversy 1764-1804 49

Part IV   Works on Black 57

Part V   Iconography 105

                Joseph Black: family tree 116

                Index 119

INTRODUCTION

Joseph Black, who was to become one of the most renowned chemists of his day, was born on 16 April 1728 in Chartrons, a suburb of Bordeaux, France. His father, an Ulsterman of Scottish origin, was a wine merchant in the city. His mother originally came from Aberdeen. The family tree is shown on page 116. Joseph went to Belfast for schooling at the age of twelve and thereafter he entered the University of Glasgow in 1744 to study the arts course followed by all young students. Four years later he commenced his medical studies, and came under the influence of, in particular, William Cullen, who had been appointed lecturer in chemistry at Glasgow in 1747. Unusually, Black performed experiments in Cullen's laboratory while studying under him.

In 1752 Black transferred to the University of Edinburgh to continue his medical studies, and there he started experiments on the causticity of alkalis. He conducted quantitative work on magnesia alba (basic magnesium carbonate) to elucidate the principle of causticity, a matter of controversy in the academic community at Edinburgh. Later Black's conclusions were to stimulate a protracted debate with the German chemist J F Meyer (see below). In 1754 Black wrote up his experiments as his thesis, which he dedicated to Cullen, and he was awarded his medical degree. The work on alkalis was extended, and Black presented a paper to the Philosophical Society of Edinburgh in which he was able to differentiate fixed air (carbon dioxide) from atmospheric air. Thus carbon dioxide was the first gas to be chemically characterised.

Black then started practising as a physician. Later that year, following the sudden decline in health of the professor of medicine and chemistry at Edinburgh, Cullen was appointed to the chair and Black was invited to succeed to the Glasgow position vacated by his former teacher. Back in Glasgow he found himself in stimulating company which included the young James Watt, who was working as the mathematical instrument maker to the University. Black took up teaching enthusiastically and developed his medical practice. But he also undertook a considerable series of experiments which resulted in the development of the concept of latent heat. Though he taught these theories to his students, he never published them, though he was frequently urged to do so by his friends.

Considerable efforts were made to arrange for Black to return to Edinburgh. This became possible in 1766 when Cullen transferred from chemistry to the chair of the institutes of medicine and Black was once again appointed to a post formerly held by Cullen. Back in Edinburgh Black's reputation as a teacher grew more illustrious and his lecture course attracted students from far and wide. In fact, those attending his course who would graduate in medicine at Edinburgh formed a rather small minority. Black developed his lecture demonstrations into a much commented-on virtuoso performance. His experimental investigations of a more fundamental nature were no longer pursued at Edinburgh though Black was active in his laboratory, responding to requests from others for advice. In particular, he helped many of those who were developing the emerging chemical industries in Scotland by analysing minerals, advising on bleaching processes, providing testimonials on the quality of iron produced by a new process and judging the economic viability of tar extraction. For many years he involved himself in a fruitless series of experiments to try to obtain alkali on a large scale basis from lime and salt. He advised the Commissioners of Police on the purity of Edinburgh's water supply and the Board of Trustees for Manufacturers on the provision of premiums for inventions.

Black fulfilled his responsibilities as a member of the Senatus Academicus of the University, and at one stage provided advice for the architect's brief when it was decided to build an impressive new college designed by Robert Adam in place of the medley of unsatisfactory buildings which had been used from the sixteenth century onwards. He also played an active role in the Royal College of Physicians of Edinburgh: he was elected President from 1788 to 1790 and helped to prepare new editions of the College's pharmacopoeia. Black acted as one of the Managers of the Royal Infirmary over a period of several years. His own medical practice was small, and probably did not extend much outside his social circle.

Though he travelled very little (only one visit to England is recorded), Black maintained contact with a large circle of scientists and other savants by correspondence. In particular, he wrote regular letters to James Watt in Birmingham and in return received intelligence of the circle which formed the Lunar Society. In Edinburgh, Black was a central figure of the brilliant group of intellectuals which had formed in the latter half of the eighteenth century. His particular friends were the geologist James Hutton and William Cullen; he also had a very close relationship with Archibald Geddes, a glass manufacturer at Leith. Black was a sociable figure, attending several dining clubs. Whenever a visitor from England or abroad came to observe the Edinburgh scene, a meal with the renowned Dr Black was much desired. Benjamin Franklin was so entertained.

Black led a well-ordered, rather unenergetic life. His teaching continued until the 1795-96 session, after which he contracted Thomas Charles Hope to take over the bulk of the lecture course. He expired in a delicate manner on 6 December 1799, and his death was widely mourned.

Black's publications

In an editorial note to Black's posthumously published *Lectures on the Elements of Chemistry*, John Robison wrote:

> Although Dr Black was duly conscious of his own merit as a chemist and a philosopher, there was never a man less assuming, or less eager to obtrude himself on the public attention. Even his discovery of fixed air, and his ingenious theory of quick-lime, would not perhaps have been committed to the press, had they not formed the subject of his inaugural dissertation, when he received his medical degree.

In fact the thesis, and the work which developed from it, was the only really significant group of publications by Black. It is known that 170 copies of *Dissertatio medica inauguralis ... De Humore acido a Cibis orto, et Magnesia Alba* (item 1) were printed.[1] The expanded version of this was presented (in English) to the Philosophical Society of Edinburgh on 5 June 1755 and it was published in the Society's journal *Essays and Observations, Physical and Literary* (item 7). It also subsequently appeared in separate, reprinted form.

The work on heat conducted at Glasgow was never published by Black, though a pirated version, probably taken from student lecture notes, appeared anonymously in 1770, as *Enquiry into the General Effects of Heat*; the publisher was John Nourse (item 108). In addition to the theory, a substantial appendix was added which deals with vessels used in chemical experiments. The work was already being quoted and adapted by others, noteworthy examples being Patrick Duguid Leslie's *A Philosophical Enquiry into the Cause of Animal Heat* of 1779 (item 221) and Adair Crawford's *Experiments and Observations on Animal Heat* of 1788 (item 242). Particularly anxious to join the bandwaggon was the communicator of intelligence (? plagiarist) J H de Magellan who wrote to James Watt with enquiries about Black's work in February

1780.[2] By May, Magellan's book *Essai sur la Nouvelle Theorie du Feu Elementaire, et de la Chaleur des Corps* (item 223) had been published. Watt wrote apologetically to Black, "I shall make no other remarks, than that he has neither done justice to you nor to the remarks I furnished him with, most of which he has suppressed as being inconsistent with the Opinion he had taken up of Wilkes having a right to discovery, merely because he printed first".[3] (Here Watt was referring to the 1771 publication on latent heat by the Swede, Johan Carl Wilcke.[4])

Watt continued to nag Black about non-publication in 1782. He wrote to say that the Swiss, Andre de Luc, and James Keir, had both asked Watt to mediate with him:

> [Keir] is going to publish a new edition of his dictionary and makes the same request that Mr De Luc does — as he must now say something on the subject of heat which he formerly declined hoping you would have done it yourself; He wishes to have his information from the fountain head and to give to Caesar those things which are Caesars — .[5]

Black replied negatively to Watt's request, saying that "I have already prepared a part of that Subject for publication and that I am resolved next summer to prepare the rest and give it to the World such as it is".[6]

Watt was not the only correspondent who tried to persuade Black to publish: Martin Wall in November 1780,[7] George Buxton in 1788[8] and Jan Ingen Housz in May 1791, all made attempts. The last pleaded his task was "to induce you to make the world know the valuable discoveries it owes to your industry and superior genius".[9] But Black, for all the efforts of his friends, never was to produce the desired work.

The only major publications subsequent to the work on alkalis were a paper communicated through Sir John Pringle to the Royal Society of London and published in 1775 (item 39), and one read to the Royal Society of Edinburgh on 4 July 1791, published in 1794 (item 62). The first paper dealt with the effect of boiling samples of water on their temperature of congelation. Black showed that unboiled water could remain liquid well below its normal freezing point. The Edinburgh paper was an analysis of water brought from the large geyser near Reykjavik in Iceland by Sir John Stanley; Black demonstrated that it contained silica held in solution by caustic alkali. His other publications are varied and slight. A number arose from others publishing the content of the letters he sent them. Indeed, John Robison commented that Black "corresponded occasionally with Seguin, and with Crell, who had been his pupil; but he did not encourage much intercourse of this kind, having found that his informations sometimes appeared in print as the investigations of the publishers"[10] (see, for example, items 46, 47, 244). Perhaps the most significant letter to be published, because of the light it throws on Black's attitude to the new chemistry being developed in France, is that to Antoine Laurent Lavoisier of 1791 (item 54).

Black's published lectures should be mentioned here, even if they were published posthumously in 1803. Their editor, John Robison, had to go through considerable agonies to order and supplement manuscript notes. In fact, only part of the end product can be considered to be Black's substantially unchanged text. The process of publication and Robison's motives have been considered in a recent paper.[11] In a sense, publication of the lectures can be seen as an act of piety, initially encouraged by Black's nephew, George Black.[12] By the time *Lectures on the Elements of Chemistry* appeared, the text must have seemed very dated. But such was Black's reputation, that editions were published in Hamburg in 1804 and 1818 (items 72 and 74) and in Philadelphia in 1806 (item 73).

Works by Black's contemporaries and the Meyer controversy

These sections contain lists of books and papers by Black's contemporaries which make reference to his work. A very broad spectrum of material is included, from critical commentaries on Black's work, to papers which Black might have himself published had he been more forthcoming. The publications generated by the dispute with Johann Friedrich Meyer (1705-1766) over the cause of causticity are listed in a separate section of their own because of the quantity of material which was generated.

Several pieces of apparatus developed by, or influenced by, Black are described or mentioned in papers published by others. Details of an experiment in which carbon dioxide was passed into caustic alkali was communicated to Francis Hutcheson at Trinity College, Dublin and in turn these particulars were forwarded to David Macbride, a physician working in Dublin who published a diagram showing the arrangement (item 207). Black probably blew some of his own thermometers (he certainly employed James Watt and Alexander Wilson to make others) and this interest is indicated in several editions of George Martine's book on the particular instrument which Black recommended to students (items 212, 241 and 249). He probably encouraged the production of medical glassware at Leith.[13] Evidence in the Edinburgh Pharmacopoeia which he edited for the Royal College of Physicians of Edinburgh (items 213, 230, 248; see also 61) and the Edinburgh and Leith Glasshouse pamphlet on Nooth's machine for making artificial mineral waters impregnated with carbon dioxide (item 239) support this suggestion. The type of portable furnace designed by Black, in use by chemists for 150 years, was described by a German student of Black in 1785 (item 227 — see also 237, 247, 250, 262 and 265), though not by Black himself. That Black strongly influenced chemistry teaching elsewhere is beyond doubt. Two printed syllabuses in this section clearly indicate this: Benjamin Rush's syllabus of 1770 for teaching in Philadelphia (item 209) and Martin Wall's of 1780 for Oxford University (item 228). A few papers indicate Black's very extensive role as a consultant to emerging Scottish industries. The cheap production of alkali sought by Black and Watt is indicated in item 219 (and is covered in detail in item 593a) and the production of saltpetre in items 220 and 226. Black's work for the Board of Manufactures is alluded to in items 261 and 266, where results of analyses of kelp from different island sources, determining alkali content, are given. These few references are an inadequate reflection of Black's activity from 1766 until his death and a much more complete picture can be obtained from the manuscript correspondence to be found in Edinburgh University Library.[14]

In 1764 J F Meyer published a theory of causticity based on the existence of an acid (*acidum pingue* or *causticum*) which combined with lime during its calcination. The acid was said to have a nature similar to that of fire. It was expansible, compressible, volatile, astringent and capable of penetrating all vessels. This was in opposition to Black's proposal which explained the nature of fixed alkali as being a compound substance of caustic alkali and fixed air (carbon dioxide). The controversy was extensively commented on and polarised positions were adopted. The bibliography includes 46 references from the period 1764 -1804. Thomas Thomson summed up the position from the 'British' point of view:

> Notwithstanding the reputation and acknowledged genius and merit of its author [Meyer's theory] never gained many followers; because the true theory of causticity which had already been published by Dr Black soon became known on the continent; and, not withstanding some opposition at first, soon carried conviction in every unprejudiced mind. Even Mr Meyer himself readily acknowledged

its truth and importance, though he did not at first, on that account, give up his own theory.[15]

Works on Black

Black and his work have never been ignored by historians of science. In the first three periods of fifty years following his death, the bibliography contains 34, 30 and 39 entries. After 1951, books and papers about Black or making extensive reference to him proliferate: well over one hundred items have been identified. However there has only been one biography in monograph form entirely devoted to Black: this is *The Life and Letters of Joseph Black*, written by the Glasgow chemist William Ramsay and published posthumously in 1918 (item 486). One of the more interesting early biographies is that written by Henry, Lord Brougham, published in 1845 (item 439). Brougham had attended one of Black's last courses of lectures, and though the tone adopted is adulatory throughout, he introduces interesting personal reminiscences of Black's style of teaching. In recent years there have been works of some considerable scholarship which include substantial material on Black. These include work on heat by Henry Guerlac (items 562, 563 and 638), an analysis of lecture notes relating to Black's lectures by Douglas McKie (items 512, 570, 572, 579, 585, 589 — see also 576) and an analysis of the state of Scottish 'philosophical' chemistry in the mid-eighteenth century by Arthur Donovan (item 609), this last being the most concentrated work to date. Recently, a symposium to mark the 250th anniversary of Black's birth was held in Edinburgh and the papers have been published (item 641).

Images of Black

Three important portraits survive: two by David Martin and one by Henry Raeburn. The earlier portrait by Martin (item 904) was painted *c*1770, shortly after Black had taken up the chair of chemistry at Edinburgh. No engravings seem to be based on it. The later Martin portrait (item 909) was commissioned by the Royal Medical Society in 1787, together with a companion work of William Cullen. This appears to be the model from which John Kay produced his caricature (item 907), also dated 1787. Kay also issued two further images of Black (items 906 and 908), the latter a double portrait with James Hutton (apart from a portrait by Raeburn, this is the only known likeness of Hutton; it is of interest that he is portrayed with his close friend). The Raeburn portrait of Black (item 910) was painted *c*1788. It has spawned at least eleven engravings. There is a further engraved portrait by Thornton (item 930).

James Tassie produced wax and glass-paste medallions (items 922-924) and Josiah Wedgwood manufactured a version of these (item 925). It was later copied by James Watt (item 927) when experimenting on his sculpturing machine after Black's death, providing a nice posthumous tribute by the engineer to his chemist friend whom he had first met fifty years earlier.

Future work

Henry Brougham pleaded 130 years ago:

> ... it must be observed that Dr Black's discoveries have been far from attaining the reputation they so well deserve as the foundation of modern chemistry; and justice to this illustrious philosopher requires that the consequences arising from his modesty and his great indifference to fame should be counteracted by a full history of his scientific labours, comparing the state of the science as he found it with that in which he left it.[16]

Today, though we may hope that a critical biography will be written which treats all aspects of Black's life, we still do not even have that more restricted "full history" which Brougham hoped for. It is hoped that this bibliography will help to stimulate a potential author.

Notes

[1] The actual number of copies is mentioned in a letter from Black to his father dated November 1755: EUL MS Gen 874, 64 no 7

[2] See Watt's reply to Magellan of 1 March 1780 in Eric Robinson and Douglas McKie *Partners in Science* (Constable: London 1970) pp 76, 77. The enquiring letter to Watt appears not to have survived.

[3] Ibid, pp 88, 89 (letter, Watt to Black, 17 May 1780)

[4] J C Wilcke 'Om Snons kyla vid Smaltningen' *Kongl. Svenska Vetenskaps Academiens Handlingar* 33 97 (1772). For additional detail, see Douglas McKie and Niels H de V Heathcote *The Discovery of Specific and Latent Heats* (Arnold: London 1935) p 78 et seq.

[5] Robinson and McKie op cit (note 2) pp 117, 118 (letter, Watt to Black, 13 December 1782)

[6] Ibid, pp 119, 120 (letter, Black to Watt, 30 January 1783)

[7] EUL MS Gen 874, vol I f 103 (letter, Watt to Black, 17 November 1788)

[8] EUL MS Gen 873, vol III f 114 (letter, Buxton to Black, 17 November 1788)

[9] EUL MS Gen 874, vol III f 205 (letter, Ingen Housz to Black, 26 May 1791)

[10] Joseph Black *Lectures on the Elements of Chemistry* vol I (Edinburgh 1803) p lxv

[11] J R R Christie in A D C Simpson (ed) *Joseph Black: a Commemorative Symposium* (Edinburgh 1982) pp 47 - 52

[12] E Robinson and D McKie *Partners in Science* (London 1970) pp 326 - 331

[13] R G W Anderson *The Playfair Collection and the Teaching of Chemistry at the University of Edinburgh 1713 - 1858* (Edinburgh 1978) p 143

[14] EUL MSS Gen 873 - 75

[15] T Thomson 'Chemistry' in *Encyclopaedia Britannica Supplement to the Third Edition* (Edinburgh 1801) p 301

[16] H P Brougham *Lives of Philosophers of the Time of George III* (London 1855) pp viii, ix

Part I

WORKS BY BLACK

Dissertatio medica inauguralis (1754)

1 Dissertatio medica inauguralis, de humore acido a cibis orto, et magnesia alba: Quam, annuente summo numine, ex auctoritate reverendi admodum viri D. Joannis Gowdie, Academiae Edinburgenae Praefecti; nec non amplissimi Senatus Academici consensu, et nobilissimae Facultatis Medicae decreto; pro gradu doctoratus, summisque in medicina honoribus et privilegiis rite et legitime consequendis, eruditorum examini subjicit Josephus Black Gallus. Ad diem II Junii, hora locoque solitis. Edinburgi: Apud G. Hamilton et J. Balfour Academiae Typographos, 1754. [6], 46 p.
Facsimiles of title page and dedication page are in *Journal of Chemical Education* 12 (1935) 225.

2 Dissertatio medica inauguralis, de humore acido a cibis orto, et magnesia alba. Quam . . . pro gradu doctoratus . . . subjicit Josephus Black . . . ad diem II. Junii . . . Edinburgi: 1754.
In: Thesaurus medicus: sive, disputationum, in Academia Edinensi, ad rem medicam pertinentium, a collegio instituto ad hoc usque tempus, delectus, a Gulielmo Smellio, S.P.E.S. habitus, tom. II, pp. 271 – 304. Edinburgi: Typis Academicis, prostant venales apud C. Elliot, et G. Creech; et Londini apud J. Murray, 1779.

3 Dissertatio medica inauguralis, de humore acido a cibis orto, et magnesia alba. Quam . . . pro gradu doctoratus . . . subjicit Josephus Black . . . ad diem II. Junii . . . Edinburgi: 1754.
In: Thesaurus medicus Edinburgensis novus: sive, dissertationum in Academia Edinensi, ad rem medicam pertinentium, ab anno 1759 ad annum 1785, delectus, ab illustri Societate Regia Medica Edinensi habitus, tom. II, pp. 271 – 304. Edinburgi & Londini: Apud

3

Works by Black

C. Elliot & G. Robinson [etc.], 1785.

4 Joseph Black's inaugural dissertation. I. Communicated by Leonard Dobbin.
Journal of Chemical Education 12 (1935) 225 – 228.
> A. Crum Brown's English translation of the earlier part of Black's *Dissertatio medica inauguralis*, dealing with "The acid humour arising from food".

5 Joseph Black's inaugural dissertation. II. Communicated by Leonard Dobbin.
Journal of Chemical Education 12 (1935) 268 – 273.
> A. Crum Brown's English translation of the later part of Black's *Dissertatio medica inauguralis*, dealing with the experiments on magnesia alba.

6 On acid humor arising from foods and on white magnesia. A translation of the Latin thesis De humore acido a cibis orto, et magnesia alba. By Joseph Black 1754. Translated by Thomas Hanson. With an introduction by Stacey B. Day. Minneapolis, Minnesota: The Bell Museum of Pathobiology, University of Minnesota Medical School, 1973. xiv, 40 p.
> Reviewed by D.F. Lardner in *British Journal for the History of Science* 7 (1974) 190 – 191.

Experiments upon magnesia alba (1756)

7 Experiments upon magnesia alba, quicklime, and some other alcaline substances.
Essays and Observations, Physical and Literary 2 (1756) 157 – 225.
> Read on 5th June 1755.

8 [Experiments upon magnesia alba. German translation by Joh. Ernst Greding]
In: Neue Versuche und Bemerkungen aus der Arztneykunst und ubrigen Gelehrsamkeit einer

Gesellschaft zu Edinburg vorgelesen und von ihr herausgegeben. Als eine Fortsetzung der medizinischen Versuche und Bemerkungen, Bd. 2, pp. 172 – 254. Altenburg, 1757.
> Not seen; reference from BUGGE, G., ed., *Das Buch der grossen Chemiker,* Bd. 1, p.242, note 7. Berlin, 1929.

9   Experiences sur la magnesie.
    Journal de Medecine, de Chirurgie et de Pharmacie 8 (1758) 254 – 261.

10 Experiments upon magnesia alba, quicklime, and some other alcaline substances.
    Essays and Observations, Physical and Literary 2nd ed. vol. 2 (1770) 172 – 248.

11 Experiences sur la magnesie blanche, la chaux vive, & sur d'autres substances alkalines, par M. Joseph Black, Docteur en Medecine.
    Observations sur la Physique, sur l'Histoire Naturelle et sur les Arts, tome 1 partie 3 (mars 1773) 210 – 220.

12 Suite des experiences de M. le Docteur Black, sur la magnesie, la chaux-vive, & sur d'autres substances alkalines.
    Observations sur la Physique, sur l'Histoire Naturelle et sur les Arts, tome 1 partie 4 (avril 1773) 261 – 275.

13 Von einsaugenden Erden, und besonders von der weissen Magnesia. Von Herrn Joseph Black.
    In: Auserlesene kleine Werke dreyer beruhmter englischer Chymisten, Herrn Priestley, Henry und Black, die Schwangerung des gemeinen Wassers mit fixer Luft, die Magnesia und Kalkerde, die faulungswidrige Kraft gewisser Arztneyen und andre erhebliche Gegenstande betreffend, pp. 133 – 152. Kopenhagen und Leipzig: Bey Johann Friedrich Heineck und Faber, 1774.

14 Experiments upon magnesia alba, quick-lime, and other alcaline substances; By Joseph Black, M.D. . . . To which is annexed, an essay on the cold produced by evaporating fluids, and of some other means of producing cold; By William Cullen, M.D. . . . Edinburgh: Printed for William Creech, Edinburgh; and for J. Murray, and Wallis and Stonehouse, London, 1777. [2], 133, [1] p.
> B.'s Experiments occupy pp. 1 – 113. The last, unnumbered, page is an advertisement for "Books printed by William Creech, for the use of students in the University of Edinburgh".

15 Experiments upon magnesia alba, quick-lime, and other alcaline substances; By Joseph Black, M.D. . . . To which is annexed, an essay on the cold produced by evaporating fluids, and of some other means of producing cold; By William Cullen, M.D. . . . Edinburgh: Printed for William Creech, Edinburgh; and for T. Cadell, London, 1782. 139 p.
> B.'s Experiments occupy pp. 3 – 116. The text proper ends on p. 135; followed by 4 pages of advertisement for "Books written by professors in the University of Edinburgh, for the use of students"; the second of these pages is numbered 137, in continuation of the pagination of the text; the fourth is dated October 1782.

16 Experiments upon magnesia alba, quick-lime, and other alcaline substances; by Joseph Black, M.D. . . . To which is annexed, an essay on the cold produced by evaporating fluids, and of some other means of producing cold; by William Cullen, M.D. . . . Edinburgh: Printed for William Creech, 1796. 134 p.

17 Experiments upon magnesia alba, quick-lime, and other alcaline substances. Edinburgh: William F. Clay; London: Simpkin, Marshall, Hamilton, Kent & Co. Ltd., 1893. 47 p.
(Alembic Club reprints, no. 1).

Brief preface by L.D. [Leonard Dobbin]. The re-issues listed below have not been seen; and there may be others.

18 Experiments upon magnesia alba, quicklime, and some other alcaline substances. Edinburgh: The Alembic Club, 1898. 46 p.
(Alembic Club reprints, no. 1).

19 Experiments upon magnesia alba, quicklime, and some other alcaline substances. Edinburgh: The Alembic Club; Chicago: University of Chicago Press, 1902. 46 p.
(Alembic Club reprints, no. 1).

20 Experiments upon magnesia alba, quicklime, and some other alcaline substances. Edinburgh: The Alembic Club; Chicago: University of Chicago Press, 1906. 46 p.
(Alembic Club reprints, no. 1).

21 Experiments upon magnesia alba, quicklime, and some other alcaline substances. Edinburgh: The Alembic Club, 1910. 46 p.
(Alembic Club reprints, no. 1).

22 Experiments upon magnesia alba, quicklime, and some other alcaline substances. Edinburgh: The Alembic Club, 1913. 46 p.
(Alembic Club reprints, no. 1).

23 KNICKERBOCKER, W.S. Classics of modern science (Copernicus to Pasteur), chap. 12, pp. 89 – 95. New York: Alfred A. Knopf, 1927.
"Joseph Black, 1728 – 1799"; a very brief biographical note followed by a reprint of two passages from pt. 1 of *Experiments upon magnesia alba, quick-lime, and other alcaline substances.*

Works by Black

24 Experiments upon magnesia alba, quicklime, and some other alcaline substances. Edinburgh: The Alembic Club, 1944. 46 p.
(Alembic Club reprints, no. 1).

25 Experiments upon magnesia alba, quicklime, and some other alcaline substances. Edinburgh: Published for the Alembic Club by E. & S. Livingstone, 1963. 46 p.
(Alembic Club reprints, no. 1).

26 LEICESTER, H.M. & KLICKSTEIN, H.S. A source book in chemistry 1400 – 1900, pp. 80 – 91. Cambridge, Massachusetts: Harvard University Press, 1965.
"Joseph Black (1728 – 1799)"; a brief biographical note followed by reprints of passages from *Experiments on magnesia alba, quicklime, and some other alcaline substances*.

27 Experiments upon magnesia alba, quick-lime, and other alcaline substances . . . New York: Readex Microprint Corp., 197–?
(Landmarks of science).
Microcard reprint of the 1777 edition. Not seen.

28 Experiments upon magnesia alba, quicklime and some other alcaline substances. New York: AMS Press, [1979?].
Reprint of the 1898 edition (Alembic Club reprints, no. 1).
Not seen; entry from *Books in Print*.

Letter to Monro (1758)

29 MONRO, A., *secundus*. Observations, anatomical and physiological, wherein Dr. Hunter's claim to some discoveries is examined, pp. 27 – 29. Edinburgh: Printed by Hamilton, Balfour & Neill, 1758.
Prints a letter from B. to Monro, dated 24th Mar. 1758, confirming that M.'s theories on the origin and use of the lymphatic veins had been formulated as early as 1755.

Works by Black

Preparations of antimony and mercury (176–?)

30 [The preparations of antimony . . .] [s.l.: s.n., 176–?] [2] p.
>No proper title: the above title is the opening words of the text.
>Apparently issued by B. to his students; 11 MSS of his lectures containing this printed table are listed by W.A. Cole (see no. 634); they range in date from 1767/68 to 1776/77. The first page of a copy in Edinburgh University Library is reproduced in no. 652, p. 105. *cf no. 31.*

31 [The preparations of mercury . . .] [s.l.: s.n., 176–?] [2] p.
>No proper title: the above title is the opening words of the text.
>Apparently issued by B. to his students; 12 MSS of his lectures containing this printed table are listed by W.A. Cole (see no. 634); they range in date from 1767/68 to 1776/77. *cf no. 30.*

32 Dr Black's table of the preparations of antimony. In: The Edinburgh new dispensatory, p. 84. Edinburgh: Printed for Charles Elliot [etc.], 1786.
>The text of the table begins: "Medicines are prepared either . . . " *cf no. 33.*
>For later editions of the *Edinburgh new dispensatory* see nos. 243, 247, 250, 258, 262.

33 [Dr Black's table of the preparations of mercury.] In: The Edinburgh new dispensatory, 4th ed., pp. 172 – 173. Edinburgh: Printed for William Creech [etc.], 1794.
>No title: the above title has been supplied by analogy with the table for antimony (no. 32); the text of the table begins: "Quicksilver is prepared for medical purposes." *cf no. 32.*
>This mercury table of B. first appears in this 4th edition, replacing Schwediauer's table in the earlier editions; for

Works by Black

later editions of the *Edinburgh new dispensatory* see nos. 258, 262.

34 [Medicamenta parantur ex antimonio vel sulphurato vel sulphure privato . . .] [s.l.: s.n., 179–?] [2] p.
> No proper title: the above title is the opening words of the text.
> A copy of this printed table is bound with John Lee's MS notes of T.C. Hope's lectures on chemistry and pharmacy, 1800 (Edinburgh University Library MS Dc.8.156, ff 2 – 3); and another with the anonymous MS notes of Black's and Hope's lectures, 1796/97 (Edinburgh University Library MS Gen.48D). *cf no. 35.*

35 [Hydrargyrus praeparatur ad usus medicos . . .] [s.l.: s.n., 179–?] [2] p.
> No proper title: the above title is the opening words of the text.
> A copy of this printed table is bound with John Lee's MS notes of T.C. Hope's lectures on chemistry and pharmacy, 1800 (Edinburgh University Library MS Dc.8.155, ff 47–48); and another with the anonymous MS notes of Black's and Hope's lectures, 1796/97 (Edinburgh University Library MS Gen.48D). *cf no. 34.*

36 [Table of the preparations of antimony, as drawn out by Dr. Black.]
In: Lectures on the elements of chemistry . . . By the late Joseph Black, M.D. . . . Now published from his manuscripts, by John Robison, LLD. . . . , vol. 2, pp. 755 – 756. Edinburgh, 1803.
> No title: the above title has been supplied by analogy with the table for mercury (no. 37); the text of the table begins: "Medicamenta parantur ex antimonio, vel sulphurato, vel sulphure privato." *cf no. 37.*
> For other editions of the *Lectures* see nos. 72 – 75.

37 Table of the preparations of mercury, as drawn out by

Dr. Black.
In: Lectures on the elements of chemistry . . . By the late Joseph Black, M.D. . . . Now published from his manuscripts, by John Robison, LL.D. . . . , vol. 2, pp. 753 – 755. Edinburgh, 1803.

> The text of the table begins: "Hydrargyrus praeparatur ad usus medicos." *cf no. 36*.
>
> For other editions of the *Lectures* see nos. 72 – 75.

Enquiry into the general effects of heat (1770)

See no. 208.

Explanation of the effect of lime (1771)

38 An explanation of the effect of lime upon alkaline salts: and a method pointed out whereby it may be used with safety and advantage in bleaching. By Joseph Black, M.D. In: HOME, F. Experiments on bleaching. By Francis Home, M.D. . . . To which are added, I. An experimental essay on the use of leys and sours in bleaching. By James Ferguson, M.D. II. An explanation of the effect of lime upon alkaline salts . . . By Joseph Black, M.D. III. An abstract of the foregoing essays, containing, practical rules and plain directions . . . , pp. 265 – 282. Dublin: Printed by T. Ewing, 1771.

> Not in the first edition, or in the French and German translations. Facsimiles of the title pages of the whole work and of B.'s contribution are in *American Dyestuffs Reporter* 44 (1955) 683, 684.
>
> "Black's section consists of a summary of his 1756 paper with a few additions, and descriptions of commercial alkalis and their caustification." – J.R. Partington in *A history of chemistry*, vol. 3, p. 142. London, 1962.

Edinburgh Pharmacopoeia (1774 – 92)

B. was involved in the preparation of the 1774, 1783 and

Works by Black

1792 editions. "Indeed, Cullen and Black were recognised abroad as largely responsible for the 1774 edition." — D.L. Cowen in ANDERSON, R.G.W. & SIMPSON, A.D.C., eds., *The early years of the Edinburgh Medical School*, p. 34. Edinburgh, 1976. See nos. 213, 230, 248 for details of these three editions.

Supposed effect of boiling (1775)

39 The supposed effect of boiling upon water, in disposing it to freeze more readily, ascertained by experiments. By Joseph Black, M.D. . . . in a letter to Sir John Pringle, Bart. P.R.S.
Philosophical Transactions, vol. 65 pt. 1 (1775) 124 – 128.
> Letter dated 11th Feb. 1775; read on 23rd Feb. 1775.

40 Effet suppose de l'ebullition sur l'eau qu'on veut glacer plus promptement, verifie par des experiences.
Observations sur la Physique, sur l'Histoire Naturelle et sur les Arts, tome 8 [no 1] (juil. 1776) 69 – 71.

41 Dr. Jos. Black's (Prof. der Chem. zu Edinb.) Erweis, dass das Wasser durch Kochen leichter friere.
Chemisches Journal 1 (1778) 194 – 196.

Letters to Smith (1777)

42 HUME, D. The life of David Hume Esq. written by himself, pp. 54 – 55, 57 – 58. London: Printed for W. Strahan; and T. Cadell, 1777.
> Prints extracts from letters from B. to Adam Smith, dated 22nd and 26th Aug. 1776, describing the last days and death of Hume. (These extracts actually appear as quotations in a letter from Smith to W. Strahan, dated 19th Nov. 1776, and printed on pp. 37 – 62).
> For different issues of the first edition of the *Life* see TODD, W.B., The first printing of Hume's Life, 1777, *Library*

5th ser. vol. 6 (1951) 123 – 125.

43 SMELLIE, W. Literary and characteristical lives of John Gregory, M.D., Henry Home, Lord Kames, David Hume, Esq. and Adam Smith, L.L.D., pp. 170 – 172. Edinburgh: Printed and sold by Alex. Smellie [etc.], 1800.
>Reprints the extracts from B.'s letters to Adam Smith dated 22nd and 26th Aug. 1776.

44 HUME, D. The letters of David Hume. Edited by J.Y.T. Greig, vol. 2, p. 449. Oxford: The Clarendon Press, 1932.
>Prints in full B.'s letter to Adam Smith, dated 26th Aug. 1776, describing Hume's last days and death; the editor states that this letter had been previously printed "with considerable alterations."

Letter to Williams (1777)

45 WILLIAMS, J. An account of some remarkable ancient ruins, lately discovered in the Highlands, and northern parts of Scotland. In a series of letters to G.C.M. Esq;, pp. 81 – 83. Edinburgh: Printed for William Creech, and sold by T. Cadell, London, 1777.
>Prints a "Letter from Dr Joseph Black, Professor of Chymistry in the university of Edinburgh, – To the Author.", dated 18th Apr. 1777, mentioning various forms of stone found in Scotland which can be "melted or softened by fire".

45a WILLIAMS, J. Letters from the Highlands of Scotland, addressed to G.C.M. Esq., pp. 38 – 40. Edinburgh: Printed for William Creech; and sold by T. Cadell, London, 1777.
>The text is the same as that of no. 45; no. 45 is in octavo while no. 45a is in quarto.

Works by Black

Letters to Crell (1783)

46 Von Herrn Professor Black in Edinburg.
Neuesten Entdeckungen in der Chemie 10 (1783) 140 – 141.
> Untitled letter in the section of the journal Auszuge aus Briefen an den Herausgeber. On vitrified sodium ammonium phosphate.

47 Von Herrn Professor Black in Edinburg.
Neuesten Entdeckungen in der Chemie 11 (1783) 97 – 99.
> Untitled letter in the section of the journal Auszuge aus Briefen an den Herausageber. On reactions of vinegar and other matters.

Letter to Graham (1783)

48 HUTCHINS, T. Experiments for ascertaining the point of mercurial congelation.
Philosophical Transactions of the Royal Society of London, vol. 73 pt. 2 (1783) *303 – *370; pl. 7.
> "This paper having been for some time mislaid could not be printed in its turn. This accounts for the double paging." – footnote on p. *303.
> Read on 10th Apr. 1783.
> Prints on pp. *305 – *306 a letter from B. to Andrew Graham, dated 5th Oct. 1779, proposing "the proper manner" of determining the freezing point of mercury.

Letter to Lind (1785)

49 CAVALLO, T. The history and practice of aerostation, pp. 31 – 33. London: Printed for the author, and sold by C. Dilly [etc.] , 1785.
> Prints a letter from B. to James Lind, dated 13th Nov. 1784, describing B.'s projected experiment with a hydrogen-filled calf's bladder, conceived in 1766.

French translation Paris: Guillot, 1786; German translation Leipzig: Schwickert, 1786.

50 RAMSAY, W. The life and letters of Joseph Black, M.D., pp. 77 – 78. London: Constable and Company Ltd., 1918.

>Again prints the letter from B. to James Lind, this time from an original MS copy; there are slight differences from Cavallo's version. Ramsay also prints Lind's letter to which B.'s was the reply (p. 77).

Directions for preparing . . . medicinal waters (1787)

See no. 239.

Remarks on . . . Cort's iron (1787)

51 A brief state of facts relative to the new method of making bar iron, with raw pit coal and grooved rollers. Discovered, and brought to perfection, by Mr. Henry Cort, of Gosport. To which is added, an appendix, containing observations of Lord Sheffield . . . and letters . . . from David Hartley, Esq; Dr. Black . . . and others, pp. 13 – 14, 16 – 17. [s.l.: s.n.], 1787.

>Prints an "Extract of a letter from Doctor Joseph Black, Professor of Chemistry at Edinburgh; dated May 15, 1786" (pp. 13 – 14); and "Extracts from Dr. Black's 'Remarks on the experiments made to prove the strength of Mr. Cort's iron;' dated Nov. 2, 1786" (pp. 16 – 17).

>A copy in the Science Museum Library, London, is bound with the manuscript *Historical account of the iron and steel manufactures, and trade*, by James Weale, private secretary to Lord Sheffield (MS 371; the *Brief state* is MS 371/3 ff. 169 – 177, 180); this copy is inscribed "from the Author [to] Lord Sheffield".

>W. identifies B.'s letter as having been addressed to Lord Stanhope, and the "Remarks" to Henry Dundas; he gives

Works by Black

>also a transcription of the original beginning and end of the "Remarks", not printed in the *Brief state* (f. 179r); and transcribes a letter from B. to Adam Jellicoe, dated 26th Feb. 1787, giving permission to publish the "Remarks", but proposing a new conclusion (f. 179) — the new conclusion was less favourable, and was not in fact printed in the *Brief state*.

52 WEBSTER, T. The case of Henry Cort, and his inventions in the manufacture of British iron. No. VII – VIII. Mechanics' Magazine, new ser. vol. 2 (1859) 85 – 86; 100 – 102.

>Reprints substantial extracts from *A brief state of facts relative to the new method of making bar iron*, 1787; in particular Extract of a letter from Doctor Joseph Black . . . and Extracts from Dr. Black's "Remarks . . . " are reprinted in full in No. VIII, pp. 101 – 102.

Spirits distilled from carrots (1790)

52a BLACK, J., HUTTON, J. & RUSSELL, J. Spirits distilled from carrots.
Transactions of the Royal Society of Edinburgh, vol. 2 pt. 1 (1790) 28 – 29 (see also p. 26).

>There is no proper title; the above title is from the margin. The authors' names are as above at the beginning of the paper, but in the order Black, Russell and Hutton at the end. A report on Hunter and Hornby's process, dated 19th May 1788 and presented to the Society on 3rd Nov. 1788; the authors had been appointed for this purpose on 3rd March 1788, as described on p. 26.

Letter to Dempster (1791)

53 SINCLAIR, J. The statistical account of Scotland, vol. 1, pp. 426 – 427, footnote. Edinburgh: Printed and sold by William Creech [etc.], 1791.

>Prints an extract from a letter from B. to George Dempster,

dated 28th Nov. 1789, describing the construction and use of a furnace for calcining marl; such a furnace was constructed by D. in the parish of Dunnichen, Forfarshire. Reprinted in *The statistical account of Scotland* [new ed., rearranged by county], vol. 13, Angus. Wakefield: EP Publications, 1976.

Letter to Lavoisier (1791)

54 Copie d'une lettre ecrite a M. Lavoisier par M. Joseph Black, Professeur en l'Universite d'Edimbourg & Associe etranger de l'Academie des Sciences de Paris.
Supplement au Journal de Paris, no 18 (1791) i – ii [Part of the Journal de Paris no 42 (11 fevr. 1791)].
> Rediscovered by DUVEEN, D. & HAHN, R., A note on some Lavoisiereana in the "Journal de Paris", *Isis* 51 (1960) 64 – 66.

55 Copie d'une lettre de M. Joseph Black . . . a M. Lavoisier.
Annales de Chimie, tome 8 [no 3] (mars 1791) 225 – 229.

56 Copy of a letter from Dr. Joseph Black, Professor of Chemistry in the University of Edinburgh, to Mr. Lavoisier at Paris.
Bee, or Literary Weekly Intelligencer 2 (1791) 39 – 40.
> "Translated from the Journal de Paris, January 19, 1791."

57 Annali di Chimica e Storia Naturale.
See no. 58.

58 GIOBERT, G.A. Extrait d'une lettre de M. Black, a M. Lavoisier.
Annales de Chimie, tome 12 [no 2] (fevr. 1792) 147.
> Forms part of the section of this issue of *Annales de Chimie* entitled Extrait du deuxieme volume des Annales Chimiques; Par M. Giobert. Pavie, 1791; G. reports that Copie d'une

Works by Black

lettre de M. Joseph Black . . . a M. Lavoisier, as published in *Annales de Chimie* 8 (1791) 225 – 229, has been published in [*Annali di Chimica e Storia Naturale*].

59 Brief vom Hrn. Prof. J. Black an Hrn. Lavoisier. Chemische Annalen 2 (1794) 35 – 38.

60 McKIE, D. Antoine Laurent Lavoisier, F.R.S. 1743 – 1794.
Notes and Records of the Royal Society of London 7 (1949) 1 – 41; pls. 1 – 2.
Reprints on pp. 39 – 41 the Copie d'une lettre . . . ; the original English letter is printed on pp. 9 – 11.

Tables of scales of heat (1792)

61 MARTINE, G. Essays on the construction and graduation of thermometers, and on the heating and cooling of bodies. A new edition, with notes and considerable additions, especially the tables of the different scales of heat, exhibited by Dr. Black, in his annual course of chemistry, pp. 183 – 186. Edinburgh: Printed for and sold by William Creech, 1792.
"Appendix. Containing tables of scales of heat." Comprises B.'s table (pp. 183 – 184); a brief account of Wedgewood's thermometer (p. 185); and W.'s table (pp. 185 – 186). In his edition of B.'s lectures (no. 71) Robison again published these: vol. 1, pp. 221 – 222 (B.'s table, slightly changed); vol. 1, pp. 222 – 226 (greatly enlarged account of W.'s thermometer); and vol. 1, pp. 226 – 227 (W.'s table, much changed).

Analysis of the waters . . . (1794)

62 An analysis of the waters of some hot springs in Iceland. Transactions of the Royal Society of Edinburgh, vol. 3 pt. 2 (1794) 95 – 126.
Read on 4th July 1791.

63 An analysis of the waters of some hot springs in Iceland; from the Transactions of the Royal Society of Edinburgh. [Edinburgh, 1794?] 34 p.

> An offprint of *Transactions of the Royal Society of Edinburgh*, vol. 3 pt. 2 (1794) pp. 95 – 126, with new pagination, etc., the addition of the word Finis at the end, and a new title page (completely transcribed above). Copy in the Watt Collection in the Science Museum Library, London.

64 An analysis of the waters of some hot springs in Iceland. By Joseph Black, M.D. &c. &c.
In: STANLEY, J.T. & BLACK, J., An account of the hot springs in Iceland; with an analysis of their waters, pp. 47 – 100. [Edinburgh?, 1794?]

> A separate issue of *Transactions of the Royal Society of Edinburgh*, vol. 3 pt. 2 (1794) pp. 95 – 153, in smaller format and type face and with rearranged contents (Stanley's "Accounts" now precede Black's "Analysis" instead of following it as in the *Transactions*). The title page gives no publisher or place or date of publication; some sources date it [1791], but the British Library and Ferguson Collection copies are on paper watermarked 1794. Black's section has its own title page, completely transcribed above.

65 Analyse des eaux de quelques sources chaudes d'Islande.
Annales de Chimie, tome 16 [no 1] (janv. 1793) 40 – 62.

66 Suite de l'analyse des eaux de quelques sources chaudes d'Islande.
Annales de Chimie, tome 17 [no 2] (mai 1793) 113 – 140.

67 Analyse des eaux de quelques sources chaudes d'Islande. Extrait.
Observations sur la Physique, sur l'Histoire Naturelle et sur les Arts, tome 43 [no 6] (dec. 1793) 457 – 459.
> Summary of nos. 65 – 66.

Works by Black

68 Zergliederung des Wassers einiger heissen Quellen auf Island.
Neues Journal der Physik, Bd. 3 Heft 1 (1796) 41 – 44.
Translated from the French summary (no. 67).

69 Beschreibung einer bequemen Methode, um kleine Quantitaten von Niederschlagen und Bodensatzen bey chemischen Untersuchungen mit Genauigkeit sammlen zu konnen.
Neues Journal der Physik, Bd. 3 Heft 1 (1796) 114 – 116.
Translation of part of B.'s paper (no. 62).

70 Beschreibung einer bequemen Methode, um kleine Quantitaten von Niederschlagen und Bodensatzen bey chemischen Untersuchungen mit Genauigkeit sammeln zu konnen.
Jahrbucher der Berg- und Huttenkunde 2 (1798) 357 – 358.
Abstracted from no. 69.

Essays . . . by Adam Smith (1795)

See no. 256.

Lectures on the elements of chemistry (1803)

71 Lectures on the elements of chemistry, delivered in the University of Edinburgh; by the late Joseph Black, M.D. . . . Now published from his manuscripts, by John Robison, LLD . . . Edinburgh: Printed by Mundell and Son, for Longman and Rees London, and William Creech Edinburgh, 1803. 2 v. lxxvi (misnumbered lxvi), [4], 556 p., 4 plates incl. frontis. port.; [2], 762 p.
In the pagination of vol. 2 161 – 168 occur twice and 177 – 184 are omitted. The three chemical plates and accompanying four-page Explanation are sometimes bound with vol. 2. In some copies only, there is an index

at the end, paginated [1] – 19.

Reviewed by Lord Brougham in the *Edinburgh Review* 3 (1803) 1 – 26 (see no. 419); reviewed also in *Annual Review* 2 (1803) 924 – 925; *British Critic* 23 (1804) 645 – 657; *Critical Review* 3rd ser. vol. 1 (1804) 86 – 95, 284 – 294; *Monthly Magazine* 16 (1803) 623 – 624; *Monthly Review* 42 (1803) 187 – 192; etc.

James Watt's copy of this work is in the Watt Collection in the Science Museum Library, London. It contains ca. 44 MS annotations or marginal marks, of which the longest is a 9-line comment at the end of Robison's dedication, signed "J Watt 1808", denying having attended B.'s lectures, while acknowledging instruction received in their private conversations; others include denials of B.'s influence on the improvement of the steam engine; some supply dates, correct figures, or make other comments on some of the experiments reported.

72 D. Josef Black's . . . Vorlesungen uber die Grundlehren der Chemie aus seiner Handschrift herausgegeben von D. Johann Robison . . . Aus dem Englischen ubersetzt und mit Anmerkungen versehen von D. Lorenz von Crell . . . Hamburg: Hoffmann, 1804. 4 v.

Not seen. Some sources give the publication date as 1804 – 1805. Crell had attended B.'s lectures in Edinburgh.

73 Lectures on the elements of chemistry, delivered in the University of Edinburgh; by the late Joseph Black, M.D. . . . Published from his manuscripts, by John Robison, LLD. . . . First American from the last London edition. Philadelphia: Printed for Mathew Carey [etc.] 1807, 1806, 1806. 3 v.

Apparently printing began in 1806, and the title pages of all three vols. would have been originally dated 1806; but printing was not completed until 1807, so to make the work appear up-to-date Carey replaced the title page of vol. 1 with a new one dated 1807. All three vols. were actually published together on ?13th Feb. 1807. For details see pp. 13 – 18 of W.D. Miles's paper in the *Library Chronicle* 22 (1956) 9 – 18. Reviewed in *Medical Repository* 2nd hexade vol. 4 (1807) 306.

74 D. Josef Black's . . . Vorlesungen uber die Grundlehren der Chemie aus seiner Handschrift herausgegeben von D. Johann Robison . . . Aus dem Englischen ubersetzt und mit Anmerkungen versehen von D. Lorenz von Crell . . . Neue wohlfeile Ausgabe. Hamburg: Bei Hoffmann und Campe, 1818. 4 v.

75 Lectures on the elements of chemistry. Philadelphia, 1827. 3 v.
    Not seen; entry from NUC.

For published (1936) extracts from T. Cochrane's MS lecture notes see no. 512; this earlier publication is superseded by McKie's full edition (no. 76).

For published (1961 – 1966) extracts from G. Cayley's MS lecture notes see nos. 570, 572, 579, 585.

For a published (1966) extract from University College MS Add. 96 see no. 585.

76 Thomas Cochrane. Notes from Doctor Black's lectures on chemistry 1767/8. Edited with an introduction by Douglas McKie. Foreword by Lord Todd of Trumpington. Wilmslow, Cheshire: Imperial Chemical Industries Limited, Pharmaceuticals Division, 1966. xxviii, 190 p.

For published (1967) extracts from the MS lecture notes of C. Blagden, N. Dimsdale, R. Dobson, H. Richardson, BM MS Add. 52495, A. Anderson, see no. 589.

77 Lectures on the elements of chemistry. New York:

AMS Press, [1978?] 3 v.
>Reprint of the 1803 edition.
>Not seen; entry from *Books in Print*.

Letters to Kames (1807)

78 TYTLER, A.F. Memoirs of the life and writings of the Honourable Henry Home of Kames . . . , vol. 2, Appendix, pp. (75) − (84). Edinburgh: Printed for William Creech; and T. Cadell and W. Davies, London, 1807.
>The article on the pages numbered (75) − (84) had been intended to follow the article ending on p.46, but was omitted in error; paginated as above, it was then inserted between pp. 75 and 76. Tytler's index reference to p. (49) is erroneous.
>Prints three letters from B. to K.: "On the attraction between clay and water" dated 23rd May 1775 (pp. (75) − (78)); "On the same subject" dated 7th Sept. 1775 (pp. (79) − (80)); "On the same subject" dated 15th Sept. 1775 (pp. (80) − (81)); and anonymous "Remarks" (pp. (82) − (84)).
>pp. 33 − 35 of *Supplement to the memoirs* . . . Edinburgh, 1809, lists the members of the Poker Club, including B.'s name.

Case of Professor Ferguson (1816)

79 MARCET, A.J.G. Case of Professor Ferguson, drawn up by Dr. Black, in May, 1797.
Medico-Chirurgical Transactions 7 (1816) 230 − 236.
>Comprises B.'s account, dated 20th May 1797; and an account of F.'s last years and death, by P. Mudie, dated 22nd Apr. 1816.

Letter to Smithson (1825)

80 A letter from Dr. Black to James Smithson, Esq.

describing a very sensible balance.
Annals of Philosophy, new ser. vol. 10 no. 1 (July 1825) 52 – 54.
> Comprises the text of B.'s letter, dated 18th Sept. 1790, plus "Note by Mr. Smithson".
> The MS draft of this letter is in Edinburgh University Library; it is addressed to James Louis Macie, Smithson's original name.

81 Dr. Black's sensible balance.
Glasgow Mechanics' Magazine 4 (1826) 154 – 155.

82 Dr. Black's sensible balance.
Quarterly Journal of Science, Literature, and the Arts 20 (1826) 161 – 162.

83 On a very sensible balance, for weighing small globules of metals. By the late Dr. Black.
Technical Repository 8 (1826) 76 – 78.

84 Balance for light weights.
Mechanics' Magazine 6 (1827) 101 – 102.

85 FARADAY, M. Chemical manipulation: being instructions to students in chemistry, pp. 62 – 64. London: Printed and published by W. Phillips [etc.], 1827.
> Reprints most of B.'s letter (no. 80).
> *Chemical manipulation* was re-issued a number of times; only 3rd edition, 1842, seen.

86 A letter from Dr. Black describing a very sensible balance. In: RHEES, W.J., ed., The scientific writings of James Smithson, pp. 117 – 120. Washington: Smithsonian Institution, 1879. (Smithsonian Miscellaneous Collections, 327 (= vol. 21 article 2)).

History of Mr. Watt's improvement . . . (1854)

87 History of Mr. Watt's improvement of the steam-engine. By Joseph Black, M.D.
In: MUIRHEAD, J.P., The origin and progress of the mechanical inventions of James Watt illustrated by his correspondence with his friends and the specifications of his patents, vol. 1, pp. xxxv – xl. London: John Murray, 1854.

88 History of Mr. Watt's improvement of the steam-engine. By Joseph Black, M.D.
In: MUIRHEAD, J.P. The life of James Watt, with selections from his correspondence, pp. 58 – 59, 95 – 96. London: John Murray, 1858.
>Reprinted from no. 87, but with the last two paragraphs omitted. 2nd ed. of the *Life*, 1859.

Autobiographical note (1918)

89 RAMSAY, W. The life and letters of Joseph Black, M.D., pp. 3 – 5. London: Constable and Company Ltd., 1918.
>First publication of the autobiographical note covering the years 1728 – 1754, taken from a copy not in B.'s hand. (p. 5 also contains another autobiographical fragment which had already been published by Ferguson (no. 423) and Robison (no. 71)).

90 GUERLAC, H. Joseph Black and fixed air: a bicentenary retrospective, with some new or little known material. Isis 48 (1957) 124 – 151.
>pp. 127 – 128 prints the autobiographical note from Edinburgh University Library MS DC.2.76(8*); this version is not in B.'s hand; and apparently differs from the version used by Ramsay (no. 89).

91 DONOVAN, A.L. The origins of pneumatic chemistry, pp. 263 – 269. (Ph.D. thesis, Princeton University,

Works by Black

1970).
> Prints extracts from Edinburgh University Library MS
> Gen. 874 Autobiographical Note; this version is B.'s original.
> *cf no. 92.*

92 DONOVAN, A.L. Philosophical chemistry in the Scottish Enlightenment, pp. 166 – 168. Edinburgh: University Press, 1975.
> Prints extracts from Edinburgh University Library MS
> Gen. 874 Autobiographical Note; this version is B.'s original.
> *cf no. 91.*

Later Publication of Correspondence (1832 –   )

Correspondence with Cullen (1832, 1859)
See nos. 433, 453.

Correspondence with Watt (1846)
See no. 444.

Correspondence with Watt (1854)
See no. 447.

Letters from Lavoisier (1871, 1887, 1921, 1943)
See nos. 460, 461, 468, 492, 526.

Letter to Smith (1899)
See no. 473a.

Correspondence (1918)
See no. 486.

Letter to Stuart (1942, 1952)
See nos. 522, 543.

Letters to Lavoisier (1949)
See no. 533.

Letter to his father (1957)
See no. 562.

Letters to A. Black (1960)
See no. 573.

Correspondence with Watt (1970)
See no. 598.

Correspondence with Beddoes (1984)
See no. 647.

Letter to the Royal Society of Edinburgh (1984)
See no. 646.

Part II

WORKS BY BLACK'S CONTEMPORARIES

201 ALSTON, C. A dissertation on quick-lime and lime-water. The second edition, with additions. Edinburgh: Printed by W. Sands, A. Murray, and J. Cochran. Sold by G. Hamilton & J. Balfour, 1754. x, 79 p.
> For Black's place in the Alston/Whytt controversy see nos. 593 and 609. For the other works by Alston and Whytt see nos. 202, 203, 204, 206.

202 ALSTON, C. A second dissertation on quick-lime and lime-water. Edinburgh: Printed by W. Sands, A. Murray, and J. Cochran. Sold by G. Hamilton & J. Balfour, 1755. vi, 64 p.

203 WHYTT, R. An essay on the virtues of lime-water in the cure of the stone. The second edition corrected, with additions . . . Edinburgh: Printed by Hamilton, Balfour and Neill, 1755. xi, 213, [1] p.
> French translation of the 2nd ed.: 1757; and nouv. ed. 1766.

204 ALSTON, C. A third dissertation on quick-lime and lime-water. Edinburgh: Printed by Sands, Donaldson, Murray, and Cochran. Sold by G. Hamilton and J. Balfour, 1757. [1], iv, 46 p.

205 NEUMANN, C. The chemical works of Caspar Neumann, M.D. . . . Abridged and methodized. With large additions, containing the later discoveries and improvements made in chemistry and the arts depending thereon, by William Lewis, M.B. . . . , pp. 473 – 474. London: Printed for W. Johnston [etc.], 1759.
> Lewis's summary of B.'s experiments on magnesia alba.

206 WHYTT, R. An essay on the virtues of lime-water and soap in the cure of the stone. The third edition. Edinburgh: Printed by Hamilton, Balfour, and Neill,

1761. xii, 220 p.
> Also issued Dublin: Printed for R. Watts and S. Watson, 1762. Reprinted in *The Works* . . . Edinburgh, 1768; German translation of the *Works* Leipzig, 1771.

207 MACBRIDE, D. Experimental essays on the following subjects: I. On the fermentation of alimentary mixtures. II. On the nature and properties of fixed air. III. On the respective powers, and manner of acting, of the different kinds of antiseptics. IV. On the scurvy; with a proposal for trying new methods to prevent or cure the same, at sea. V. On the dissolvent power of quick-lime. London: Printed for A. Millar, 1764. xiii, [2], 267 p.; 2 folded tables; 4 plates.
> Essay II is on pp. 25 – 107; the plate facing p.52 illustrates an apparatus attributed to B. by Macbride.
> French translation, 1766; German translation, 1766; later English editions under the title *Experimental essays on medical and philosophical subjects*: 2nd ed., 1767; 3rd ed., 1776.

208 An enquiry into the general effects of heat; with observations on the theories of mixture . . . With an appendix on the form and use of the principal vessels . . . London: Printed for J. Nourse, 1770. viii, 119 p.
> A pirated edition of B.'s lectures on heat and chemical apparatus.

209 [RUSH, B.] Syllabus of a course of lectures on chemistry. Philadelphia, 1770. 48 p.
> "This Syllabus, although not compiled by Black and not bearing his name, was the first printed record of the Scottish chemist's course." – W.D. Miles in *Library Chronicle* 22 (1956) 11; Miles discusses the extant copies of Rush's *Syllabus*, and the extant MS notes of Rush's chemistry lectures, in *Library Chronicle* 22 (1956) 9 – 18 – see especially pp. 9 – 11, 17 – 18; and in *Chymia* 4 (1953) 37 – 77 – see especially pp. 44 – 55.

Facsimile of the title page in SMITH, E.F., *Old chemistries*, p. 12. New York; London: McGraw-Hill Book Company, Inc., 1927.
Facsimile reprint, with an introduction by L.H. Butterfield: Philadelphia: Friends of the University of Pennsylvania Library, 1954. 15, 48 p.

210 MACQUER, P.J. A dictionary of chemistry . . . Translated from the French. With notes and additions by the translator. London: Printed for S. Bladon, 1771.
> The translator [J. Keir] added "notes and new articles, particularly on the recent works of Black and Cavendish."
> — E.L. Scott in the *Dictionary of Scientific Biography*, vol. 7 (1973) 277.

211 Experiences du Docteur Black, sur la marche de la chaleur dans certaines circonstances.
Observations sur la Physique, sur l'Histoire Naturelle et sur les Arts (sept. 1772) 156 – 166.
> Not seen; this early duodecimo series of Rozier's *Observations* was published in 18 monthly issues from July 1771 to Dec. 1772, preceding tome 1 (1773) of the quarto series; becoming scarce it was reprinted in 1777 in two quarto volumes entitled *Introduction aux Observations* . . . Detailed bibliographical information on these *Observations*, and a reprint of the 1772 *Experiences* are given by D. McKie in *Annals of Science* 13 (1957) 73 – 89 (no. 565).
> An account of six experiments sent from Edinburgh "par un des Disciples du Docteur Black."
> For the quarto reprint see no. 218; and for a German translation no. 229.

212 MARTINE, G. Essays and observations on the construction and graduation of thermometers, and on the heating and cooling of bodies. The second edition. Edinburgh: Printed by Alexander Donaldson, 1772.
vi, 177 p.; plate.

Dedication: "To Dr. Joseph Black, Professor of Medicine and Chemistry in the University of Edinburgh, this new edition of Dr. Martine's essays is dedicated by the editor"; the same dedication appears in the 3rd ed., 1780, and the 4th ed., 1787; for the 1792 ed. see no. 249.

The plate is a comparison of 15 thermometer scales; it was first published in Martine's *Essays medical and philosophical*, 1740; then re-used in his *Essays and observations*, 1772, 1780 & 1787; and then re-engraved on a larger scale for the 1792 ed. Copies were possibly issued by B. to his students; a copy (of the original version) is bound with Beaufoy's MS notes of B.'s lectures, St. Andrews University.

213 Pharmacopoeia Collegii Regii Medicorum Edinburgensis. Edinburgi: Apud G. Drummond et J. Bell, 1774. 2, 184, 16 p. [Sixth edition]

B. served on the Revision Committee; see no. 611 for details.

214 LAVOISIER, A.L. Opuscules physiques et chymiques, tome 1, premiere partie, pp. 1 – 184. Paris: Chez Durand, Didot, Esprit, 1774.

The *premiere partie* is entitled "Precis historique sur les emanations elastiques qui se degagent des corps . . . ". In particular chap. 7, pp. 37 – 43, "Theorie de M. Black sur l'air fixe ou fixe . . . "; chap. 9, pp. 47 – 56, "Application de la doctrine de M. Black . . . par M. Macbride"; chap. 11, pp. 59 – 65, "Theorie de M. Meyer . . . "; chap. 12, pp. 65 – 71, "Developpement de la theorie de M. Black . . . par M. Jacquin"; chap. 13, pp. 71 – 86, "Refutation de la theorie de Messieurs Black, Macbride & Jacquin, par M. Crans".

Analysed in *Observations sur la Physique, sur l'Histoire Naturelle et sur les Arts* 3 (1774) 152 – 157; and [by Lavoisier himself] in *Histoire de l'Academie Royale des Sciences* (annee 1774, published 1778) 71 – 78.

The *Opuscules* are reprinted in *Oeuvres de Lavoisier*, tome 1, pp. 437 – 655. Paris: Imprimerie Imperiale, 1864. In

Works by Black's Contemporaries

particular the *premiere partie* is reprinted on pp. 443 — 555, and chaps. 7 — 13 on pp. 468 — 498.

215 BUCQUET, J.B.M. Experiences physico-chimiques sur l'air qui se degage des corps dans le temps de leur decomposition & qu'on connoit sous le nom vulgaire d'air fixe. Memoires de Mathematique et de Physique Presentes a l'Academie Royale des Sciences, par Divers Savans, & Lus dans ses Assemblees 7 (1776) 1 — 17.

Read 24th Apr. 1773.
Includes Bucquet's improvements to Macbride's apparatus for generating carbon dioxide.
A report on this paper, by Desmarets and Lavoisier, is published in *Oeuvres de Lavoisier*, tome 4, pp. 155 — 158.
Paris: Imprimerie Imperiale, 1868.

216 HOME, H., Lord Kames. The gentleman farmer. Being an attempt to improve agriculture, by subjecting it to the test of rational principles. Edinburgh: Printed for W. Creech, Edinburgh, and T. Cadell, London, 1776. xxvi, 409 p.; 3 plates.

"An imprimatur from one of the ablest chymists of the present age [Black], has given me some confidence of being in the right tract." — pp. xii — xiii of the Preface.
Also issued Dublin: Printed by James Williams, 1779.

217 LAVOISIER, A.L. Essays physical and chemical . . . Translated from the French, with notes, and an appendix, by Thomas Henry. London: Printed for Joseph Johnson, 1776.

Reprinted, with a new introduction by F. Greenaway, London: Frank Cass & Co. Ltd., 1970.

218 Experiences du Docteur Black, sur la marche de la chaleur dans certaines circonstances. Introduction aux Observations sur la Physique, sur l'Histoire Naturelle et sur les Arts 2 (1777) 428 — 431.

Reprinted from no. 211; the reprint is exact except for the misprint of 211 for 212 in the comments on the fifth experiment.

219 [Manufacture of acid and alkali from sea salt]
Journals of the House of Commons 37 (1778/1780) 865 etc.
> Petition of A. Fordyce (pp. 865 – 866); petition of J. Keir (p. 891); petition of P.T. De Bruges (p. 892); petition of James Watt on behalf of himself and Black (pp. 892 – 893); report of committee on the petition of Fordyce (pp. 893 – 896); petition of J. Collison (p. 897); petition of J. Fry (p. 908); petition of R. Shannon (pp. 909 – 910); petition of S. Garbett (p. 912); report of committee on the petitions of Keir and De Bruges (pp. 913 – 915); petition of I. Cookson (pp. 916 – 917); bill to remove duties on salt used to prepare marine acid and fixed fossil or mineral alkali, ordered (p. 917); presented and read (p. 921); report of committee on the petitions of Watt & Black, Collison, and Fry (pp. 929 – 930).

220 FISCHER, P. Eine neue Art die Salpeternaphte zu bereiten.
Neue Philosophische Abhandlungen der Baierischen Akademie der Wissenschaften 1 (1778) 389 – 398.

220a LAMBERT, J.H. Beobachtungen uber die Dinte und das Papier, von Hrn. Lambert.
Chemisches Journal 1 (1778) 224 – 229.
> From *Nouveaux Memoires de l'Academie Royale des Sciences et Belles-Lettres* (for 1770, published 1772) 58 – 67.
> The editor of the journal, Crell, adds a footnote (p. 229): B., among others, advocates adding fine coal-dust to ink to increase its permanence.

221 LESLIE, P.D. A philosophical inquiry into the cause of animal heat. With incidental observations on several

phisiological and chymical questions, connected with this subject. London: Printed for S. Crowder [etc.], 1778. viii, 362 p.

> ". . . vigorously defended Black's hypothesis . . . against the recent claims of A.L. Lavoisier and Joseph Priestley, in an expanded version of his medical dissertation." – C.E. Perrin in *Annals of Science* 40 (1983) 125. [The medical dissertation was *De caloris animalium causa*. Edinburgh, 1775.]

221a NAIRNE, E. E. Nairne's Versuche, ob das geschmolzene Eis vom Seewasser suss sey – welches seine specifische Schwere sey – und bey welchem Grade der Kalte das Seewasser anfange zu frieren.
Chemisches Journal 1 (1778) 212 – 214.

> From *Philosophical Transactions of the Royal Society of London* 66 (1776) 249 – 256.
> The editor of the journal, Crell, adds a footnote (p. 214): B. had told him in 1769 of his researches showing that undisturbed water can be cooled $10°$ below freezing point, but rises these $10°$ on turning to ice.

222 CLEGHORN, W. Disputatio physica inauguralis, theoriam ignis complectens. Quam . . . pro gradu doctoratus . . . subjicit Gulielmus Cleghorn . . . ad diem 12. Septembris . . . Edinburgi: Apud Balfour et Smellie, 1779. 59 p.

> In his edition of T. Cochrane's lecture notes (pp. xvii – xviii) D. McKie refers to pp. 12 – 15 of this work, stating that Cleghorn borrowed some material directly from B., translating it into Latin and presenting it as his own.
> Reprinted in *Thesaurus medicus Edinburgensis novus*, tom. 4. Edinburgi & Londini, 1785.
> Facsimile of the title page, reprint of the Latin text, and an English translation, in *Annals of Science* 14 (1958) 1 – 82.

222a CRAWFORD, A. Experiments and observations on animal heat, and the inflammation of combustible bodies.

London: J. Murray; J. Sewell, 1779. 4 p.l., 120 p.

223 MAGELLAN, J.H. de. Essai sur la nouvelle theorie du feu elementaire, et de la chaleur des corps . . . Londres: De l'Imprimerie de W. Richardson, 1780.
> Part 4 of the author's *Collection de differens traites sur des instrumens d'astronomie.*

224 MARTINE, G. Essays and observations on the construction and graduation of thermometers. 1780. See no. 212.

225 WALL, J. Medical tracts by the late John Wall, M.D. of Worcester. Collected and republished by Martin Wall, M.D., pp. 245 – 247. Oxford: Printed: Sold by D. Prince and J. Cooke, Oxford; and T. Cadell, London, 1780.
> Martin Wall states that he attended B.'s lectures in 1767, at which time B. explained the method of impregnating water with fixed air.

225a LANDRIANI, M. Del calor latente dissertazione. In his: Opuscoli fisico-chimici, pp. 81 – 149. Milano: Nelle Stampe di Gaetano Pirola, 1781.
> In particular pp. 98 – 104 for B.'s researches. This *dissertazione* is translated into French in *Observations sur la Physique, sur l'Histoire Naturelle et sur les Arts* 26 (1785) 88 – 100; 197 – 207.

226 H., J.G. Salpeternaphtha nach der Fischerschen Methode. Neuesten Entdeckungen in der Chemie 5 (1782) 51 – 69.

227 REUSS, A.C. Beschreibung eines neuen chemischen Ofens. Leipzig: Bei Christian Gottlob Hilschern, 1782. 75 p.; 3 plates.
> Reviewed by Crell in *Neuesten Entdeckungen in der Chemie* 7 (1782) 205 – 207. R. attended B.'s lectures and provided this account, the most comprehensive contemporary

description of any of B.'s apparatus.

228 WALL, M. A syllabus of a course of lectures in chemistry, read at the Museum, Oxford. By Martin Wall, M.D., Public Reader in Chemistry. Oxford: Printed for D. Prince and J. Cooke, 1782.
> Lectures 7 – 8, pp. 8 – 10, "Of the chemical apparatus"; similar to B.'s order and details. W. attended B.'s lectures (see no. 225).

229 Blacks Erfahrungen uber den Gang der Warme unter gewissen Umstanden.
Neuesten Entdeckungen in der Chemie 9 (1783) 218 – 224.
> Translated from no. 211/218.

230 Pharmacopoeia Collegii Regii Medicorum Edinburgensis. Edinburgi: Apud Joannem Bell; et Londini, Apud Geo. Robinson, 1783. xvi, 236, [17] p. [Seventh edition]
> See note to no. 213.

231 BLAGDEN, C. History of the congelation of quicksilver. Philosophical Transactions of the Royal Society of London, vol. 73 pt. 2 (1783) 329 – 397.
> Refers to B. on pp. 344 – 345 and pp. 349 – 351.

232 CAVENDISH, H. Observations on Mr. Hutchins's experiments for determining the degree of cold at which quicksilver freezes.
Philosophical Transactions of the Royal Society of London, vol. 73 pt. 2 (1783) 303 – 328.
> Refers to B. on pp. 309 – 310 and pp. 312 – 313 footnote.

233 LAVOISIER, A.L. Physikalisch-chemische Schriften. Aus dem Franzosischen ubersetzt von Christ. Ehrenfr. Weigel. Erster Band. Greifswald: Bey Anton Ferdinand Rose, 1783.

Works by Black's Contemporaries

234 DALRYMPLE, J. Address and proposals from Sir John Dalrymple, Bart. on the subject of the coal, tar, and iron, branches of trade. Edinburgh printed; London reprinted, 1784. 15 p.
> pp. 6 – 7 refer to B.'s approval of Lord Dundonald's process for extracting tar and H. Cort's process for making bar-iron.

235 LUBBOCK, R. Dissertatio physico-chemica, inauguralis, de principio sorbili, sive communi mutationum chemicarum causa, quaestionem, an phlogiston sit substantia, an qualitas, agitans; et alteram ignis theoriam complectens: Quam . . . pro gradu doctoris . . . subjicit Ricardus Lubbock . . . ad diem 13. Septemb. . . . Edinburgi: Apud Balfour et Smellie, 1784. [6], 137 p.
> The second footnote on p. 12 has been used as evidence for B.'s early conversion to the antiphlogistic theory; refuted by C.E. Perrin (no. 640).

236 CRELL, L. Ueber die kurzeste Bezeichnungsart der Korper, und ihrer Veranderungen, bey doppelten Verwandtschaften.
Chemische Annalen (1785 Bd. 1) 346 – 349; pl. fig 6.

237 The Edinburgh new dispensatory . . . With . . . new tables of elective attractions, single and double; of antimony, mercury, &c. and copperplates of chemical and pharmaceutical instruments . . . The whole being an improvement upon the New dispensatory of Dr Lewis. By gentlemen of the faculty at Edinburgh. Edinburgh: Printed for Charles Elliot, Edinburgh; and G.G.J. and J. Robinson, London, 1786. [2], xxxii, 720 p.; 3 plates.
> B.'s furnace is described and illustrated on pp. 49 – 52 and pl. 1 figs. 7 – 10; a footnote on p. 50 states that such furnaces can be obtained from John Sibbald, who makes them "under the immediate inspection of Dr Black". B.'s still is described and illustrated on p. 57 and pl. 3 fig. 4. A still boiler possibly used by B. is in the Royal Museum of

Works by Black's Contemporaries

Scotland (see no. 619, pp. 110 – 111).

238 LUC, J.A. de. Idees sur la meteorologie. Londres: De l'Imprimerie de T. Spilsbury . . . se vend chez P. Elmsly . . . Londres; et chez la Veuve Duchesne . . . Paris, 1786 – 1787. 2 v.

> The main text of tome 1 was made available to de Luc's friends before publication; the comments he received formed the basis of an "Appendice a ce premier volume", and tome 1 as published incorporates this appendix (it follows p. 543 of the main text, but is paginated 485 – 516; Robison is wrong in stating the appendix is in tome 2.)
> B.'s work was discussed on pp. 177 – 179, 214 – 215; Appendice pp. 503 – 514 contain J. Watt's comments on these passages, an anonymous comment, and de Luc's reactions to these comments.
> Re-issued Paris: La Veuve Duchesne, 1787 (with the preliminary leaves of tome 1 pt. 1 reprinted and with the title page changed from 1786).

239 Directions for preparing aerated medicinal waters, by means of the improved glass machines made at Leith glass-works. Edinburgh: Printed for William Creech, 1787. 12, [1] p.; 1 plate.

> The Duveen Collection copy of this work was formerly in the library of Sir John Hall "who has written on the titlepage 'Suppd. by Dr. Black.' In the preface to Black's 'Lectures on the elements of chemistry,' the editor (Robison) mentions that Black was intimate with the manager of the Leith glass works." – D.I. Duveen in *Bibliotheca alchemica et chemica*, p. 81. London, 1949. The Duveen Collection is now in the University of Wisconsin Libraries; and *Directions* . . . is attributed to B. in NEU, J., ed., *Chemical, medical and pharmaceutical books printed before 1800*, item 1170, p. 72. Madison: University of Wisconsin Press, 1965.

240 Vom Hrn. Dr. Dollfuss in London. Chemische Annalen (1787 Bd. 2) 60 – 61.

Untitled letter in the section of the journal Vermischte chemische Bemerckungen, aus Briefen an den Herausgeber. Mentions that Beddoes and Black are not indisposed to accept Lavoisier's theory.

241 MARTINE, G. Essays and observations on the construction and graduation of thermometers. 1787. See no. 212.

242 CRAWFORD, A. Experiments and observations on animal heat, and the inflammation of combustible bodies. The second edition. London: Printed for J. Johnson, 1788. [16], 491 p.; 4 plates.
> Among the references to B., on pp. 72 – 74 is an account of his experiments on the melting of ice "that appears to be wholly imaginary." — McKie and Heathcote in *The discovery of specific and latent heats*, p. 40. London, 1935. (This account does not appear in the first edition of Crawford's work, 1779.)

243 The Edinburgh new dispensatory. 2nd ed., 1789, not seen.

244 SEGUIN, A. Observations generales sur le calorique & ses differens effets, & reflexions sur la theorie de MM. Black, Crawfort, Lavoisier & de Laplace, sur la chaleur animale & sur celle qui se degage pendant la combustion; avec un resume de tout ce qui a ete fait & ecrit jusqu'a ce moment sur ce sujet.
Annales de Chimie 3 (1789) 148 – 242.
> For a German translation see no. 252.

245 WIEGLEB, J.C. A general system of chemistry, theoretical and practical . . . Taken chiefly from the German of M. Wiegleb. By C.R. Hopson, pp. 4S4r – 4T1r; pl. 1 figs. 9 – 12. London: Printed for G.G.J. and J. Robinson, 1789.
> "Dr. Black's portable furnace, from the Edinburgh New

Dispensary".

Reprinted in microfilm in *The eighteenth century*, reel 452, item 1. Woodbridge, Conn.: Research Publications, Inc., 1984.

246 FOURCROY, A.F. de. Elements of natural history and chemistry . . . Translated from the last Paris edition, 1789, being the third . . . London: Printed for C. Elliot and T. Kay . . . and C. Elliot, Edinburgh, 1790. 3 v.

> This and the 4th edition of R. Heron's translation are dedicated: "To Dr. Black. This translation of the last edition of a book which he has recommended to students of chemistry, is respectfully dedicated by the translator".

247 The Edinburgh new dispensatory . . . With new tables of elective attractions, of antimony, of mercury, &c. and six copperplates of the most convenient furnaces, and principal pharmaceutical instruments. Being an improvement of the New dispensatory by Dr Lewis. The third edition; With . . . a full and clear account of the new chemical doctrines published by Mr. Lavoisier. Edinburgh: Printed for William Creech. And sold in London by G.G. and J.J. Robinsons, and T. Kay, 1791. 665 p.; 3 plates.

> This edition was edited by Andrew Duncan. B.'s furnace is again described and illustrated, on pp. 81 − 84 and pl. 1 figs. 7 − 10; and his still on pp. 89 − 90 and pl. 3 fig. 4.

248 Pharmacopoeia Collegii Regii Medicorum Edinburgensis. Edinburgi: Apud Bell & Bradfute; et Londini, Apud G.G.J. & J. Robinson, 1792. xix, 254 p. [Eighth edition]

> See note to no. 213.

249 MARTINE, G. Essays on the construction and graduation of thermometers, and on the heating and cooling of bodies. By George Martine, M.D. A new edition, with

notes and considerable additions, especially the tables of the different scales of heat, exhibited by Dr. Black, in his annual course of chemistry. Edinburgh: Printed for and sold by William Creech, 1792. vi, 186 p.; plate.

> The Advertisement on p. iii states " . . . this book is recommended by Dr. Black, to the students attending his class . . . "
> For the "tables of the different scales of heat" see no. 61.

250 The Edinburgh new dispensatory . . . With new tables of elective attractions of antimonial and mercurial preparations, &c. and several copperplates of the most convenient furnaces . . . Being an improvement of the New dispensatory by Dr Lewis. The fourth edition; With . . . a full and clear account of the new chemical doctrines published by Mr. Lavoisier. Edinburgh: Printed for William Creech. And sold in London by G.G. and J.J. Robinsons, and T. Kay, 1794. xxxii, 622, [1] p.; 3 plates.

> Dedication by the editor, John Rotheram: "To Joseph Black, M.D.", mentioning that B. recommends the dispensatory in his lectures, and that it is designed to be as useful as possible to B.'s students.
> B.'s furnace is again described and illustrated, on pp. 47 – 50 and pl. 1 figs. 7 – 10; on p. [623] it is stated that such furnaces can be obtained from Ebenezer Allan.

251 BEDDOES, T. Letters from Dr. Withering, of Birmingham, Dr. Ewart . . . Dr. Thornton . . . and Dr. Biggs . . . together with some other papers, supplementary to two publications on asthma, consumption, fever, and other diseases. Bristol: Printed by Bulgin and Rosser [etc.], [1794].

> Contains an introductory dedicatory letter "To Joseph Black, M.D. Professor of Chemistry in the University of Edinburgh."

252 SEGUIN, A. Allgemeine Bemerkungen uber den

Warmestoff, und seine verschiedenen Wurkungen; und Bemerkungen uber die Theorie von Black, Crawford, Lavoisier und de la Place, uber die thierische Warme, und die Verbrennung, u.s.w.
Beytrage zu den Chemischen Annalen 5 (1794) 70 – 94.
> Translated from no. 244.

253 STANLEY, J.T. An account of the hot springs near Rykum in Iceland: in a letter to Dr Black from John Thomas Stanley, Esq.
Transactions of the Royal Society of Edinburgh, vol. 3 pt. 2 (1794) 127 – 137.
> Letter dated 30th Mar. 1792; read on 30th Apr. 1792.
> This paper appeared in the same issue of the *Transactions* as B.'s paper An analysis of the waters of some hot springs in Iceland; and was also separately issued together with B.'s paper in a volume entitled *An account of the hot springs in Iceland; with an analysis of their waters* (no. 64).

254 STANLEY, J.T. An account of the hot springs near Haukadal in Iceland: In a second letter to Dr Black from John Thomas Stanley, Esq.
Transactions of the Royal Society of Edinburgh, vol. 3 pt. 2 (1794) 138 – 153.
> Letter dated 15th Aug. 1791; read on 7th Nov. 1791.
> This paper appeared in the same issue of the *Transactions* as B.'s paper An analysis of the waters of some hot springs in Iceland; and was also separately issued together with B.'s paper in a volume entitled *An account of the hot springs in Iceland; with an analysis of their waters* (no. 64).

255 FOURCROY, A.F. de. Chimie.
Encyclopedie Methodique: Chimie, Pharmacie et Metallurgie 3 (An 4 = 1795/96) 262 – 781.
> In particular pp. 362 – 389 "Veritable decouverte de l'air fixe & des gaz; distinction & caracteres des principaux fluides elastiques; elle s'etend pour les dates de 1750 a 1774, & pour les hommes depuis Venel & Black jusqu'a

Lavoisier". Fourcroy states on p. 715 that he is writing in Mar. 1797.

256 SMITH, A. Essays on philosophical subjects. By the late Adam Smith, LL.D. . . . To which is prefixed, an account of the life and writings of the author; by Dugald Stewart, F.R.S.E. London: Printed for T. Cadell Jun. and W. Davies . . . ; and W. Creech, Edinburgh, 1795. xcv, 244 p.
"Advertisement by the editors" signed by Joseph Black, James Hutton.
This work has been reprinted many times, most recently as vol. 3 of the *Glasgow edition of the Works and Correspondence of Adam Smith*. Oxford: Clarendon Press, 1980; see pp. 28 – 30 of the latter for a bibliographical note listing nine editions between 1795 and 1967.

257 FOURCROY, A.F. de. Elements of chemistry, and natural history. To which is prefixed The philosophy of chemistry. By A.F. Fourcroy. Translated from the fourth and last edition of the original French work, by R. Heron. London: Printed for J. Murray and S. Highley [etc.] 1796. 4 v.

258 The Edinburgh new dispensatory.
5th ed., 1797, not seen.

259 FAUJAS DE SAINT–FOND, B. Voyage en Angleterre, en Ecosse et aux Iles Hebrides, tome 2, pp. 267 – 272. Paris: H.J. Jansen, 1797.
"brief, uninspired account of a visit to Black in 1784 (largely devoted to a description of Black's portable furnace)."
– H. Guerlac in *Dictionary of Scientific Biography* 2 (1970) p. 182 col. 2.
English translations: 1799, 1809 (in MAVOR, W.F., *The British tourists* . . .), 1907.

260 KLAPROTH, M.H. Beitrage zur chemischen Kenntniss

der Mineralkorper, Bd. 2, pp. 99 – 108. Posen: Bei Decker und Compagnie; Berlin: Bei Heinrich August Rottmann, 1797.
> "Chemische Untersuchung des Wassers der siedenden Quelle, zu Reikum auf Island"; in particular pp. 105 – 108 for a comparison with B.'s researches.

260a TYTLER, J. Chemistry.
Encyclopaedia Britannica, 3rd ed., vol. 4 (1797) 374 – 635.
> In the Preface in vol. 1 the proprietors state (p. xv) " . . . CHEMISTRY [etc. etc.] we have reason to believe were compiled by Mr James Tytler chemist . . . "; and later (p. xvi) "There is, however, no man to whom the Proprietors . . . feel themselves under greater obligations than to Dr Black, for the very handsome offer which he made to the person who was at first entrusted with the chemical department of the Work." References to Black in the article Chemistry can be found from its index (vol. 4, p. 620, col. 1); these include an account of his furnace (pp. 452 – 453; pl. CXXXIII). See also no. 419a.

261 JAMESON, R. An outline of the mineralogy of the Shetland Islands, and of the Island of Arran . . . With an appendix; containing observations on peat, kelp, and coal. Edinburgh: Printed for William Creech; and T. Cadell, Jun. and W. Davies, London, 1798.
> pp. 193 – 196 contain J.'s analyses of kelps, some by Kirwan's method, some by B.'s.

262 The Edinburgh new dispensatory . . . With new tables of elective attractions, of antimonial and mercurial preparations, &c. and several copperplates of the most convenient furnaces . . . Being an improvement of the New dispensatory by Dr Lewis. The sixth edition; With a full and clear account of the new chemical doctrines published by M. Lavoisier. Edinburgh: Printed for William Creech; and sold in London by G.G. &

J. Robinson, T. Kay, and T. Cox, 1801. xxxii, 622, [1] p.; 3 plates.

> Same (see no. 250) dedication by Rotheram, this time with the added heading "Dedication to the late Dr Black". B.'s furnace is again described and illustrated on pp. 47 – 50 and pl. 1 figs. 7 – 10; the advertisement on p. [623] now gives the maker's name as Ebenezer Annan.

263 LAVOISIER, A.L. Opuscules physiques et chimiques. Seconde edition. Paris: Chez Deterville, 1801.
> Not seen; entry from Duveen and Klickstein. There are two distinct issues.

264 [SCHERER, A.N.] Black's Lothrohr. Allgemeines Journal der Chemie 8 (1802) 575 – 576; pl. 5 fig. 3.
> This issue of the journal is numbered according to three systems: 11th Heft of the 4th Jahrgang; 5th Heft of the 8th Band; and Heft 47 [of the whole set].
> Description of the blowpipe from Jacquin.

265 KNIGHT (RICHARD & GEORGE). Knight's catalogue of chemical apparatus, &c., pp. 3 – 4; 5; 11. London, [180–?]
> "Knight's improved, universal, portable furnace, on the principle of Dr. Black . . . " (pp. 3 – 4).
> "Copper stills and worm tubs . . . adapted for the Black's furnace" (p. 5).
> "Dr. Black's blow pipe . . . " (p. 11).

266 ACCUM, F. Description of the portable furnace constructed by Dr. Black, and since improved. In a letter from Mr. Accum. Journal of Natural Philosophy, Chemistry, and the Arts 6 (1803) 273 – 275.
> Already described by Accum in his *A system of theoretical and practical chemistry*, vol. 2, p. 357; pl. 4, fig. 1. London, 1803. Accounts of B.'s furnaces continued to appear in the

Works by Black's Contemporaries

early 19th century, e.g. in REES, A., *The cyclopaedia*, text-vol. 15 under "Furnaces"; and plate-vol. 2 under "Furnaces Place II". London, 1819 – 1820.

267 FYFE, A. Essay upon the comparative value of kelp and barilla, founded upon accurate experiments. Prize Essays and Transactions of the Highland Society of Scotland 5 (1820) 10 – 64.

pp. 21 – 30, 36 – 37, describe the analysis of kelps and barilla for alkali content according to "the method proposed by Dr. Black."

Part III

THE BLACK/MEYER CONTROVERSY 1764 – 1804

301 MEYER, J.F. Chymische Versuche, zur naheren Erkenntniss des ungeloschten Kalchs, der elastischen und electrischen Materie, des allerreinsten Feuerwesens, und der ursprunglichen allgemeinen Saure. Nebst einem Anhange von den Elementen. Hannover und Leipzig: Bey Johann Wilhelm Schmidt, 1764. [24], 418, [34] p.

302 MEYER, J.F. Essais de chymie, sur la chaux vive, la matiere elastique et electrique, le feu, et l'acide universel primitif . . . Traduits de l'allemand . . . par P.F. Dreux. Paris: G. Cavelier, 1766. 2 v.

303 WIEGLEB, J.C. Kleine chymische Abhandlungen von dem grossen Nutzen der Erkenntniss des Acidi pinguis bey der Erklarung vieler chymischen Erscheinungen . . . nebst einer Vorrede worinnen Herrn Meyers Leben erzahlt und von dessen Verdiensten gehandelt wird von E.G. Baldinger. Langensalza, In Joh. Christian Martini Verlag, 1767. 112 p.

304 BOEHM, M.F. Examen acidi pinguis, sub praesidio Dn. Jacobi Reinboldi Spielmann . . . disquisitioni subjiciet Michael Fridericus Boehm auctor . . . Argentorati: Typis Johannis Henrici Heitzii, 1769. 38 p.

305 JACQUIN, N.J. Examen chemicum doctrinae Meyerianae de acido pingui, et Blackianae de aere fixo, respectu calcis. Vindobonae: Apud Joannem Paulum Kraus, 1769. 96 p.

306 CRANTZ, H.J.N. von. Examinis chemici doctrinae Meyerianae de acido pingui et Blackianae de aere fixo respectu calcis rectificatio. Lipsiae: Impensis Ioannis Pauli Kraus, 1770. 212 p.

307 JACQUIN, N.J. Chymische Untersuchung der

Black/Meyer Controversy

Meyerischen Lehre von der fetten Saure . . . Wien, 1770.

308 MEYER, J.F. Chymische Versuche zur naheren Erkenntniss des ungeloschten Kalchs . . . Zweite . . . Ausgabe. Hannover: Schmidt, 1770. 418, 48 p.

309 WIEGLEB, J.C. Vertheidigung der Meyerischen Lehre vom Acido pingui, gegen verschiedene darwider gemachte Einwurfe. Altenburg: In der Richterischen Buchhandlung, 1770. 115 p.; errata leaf.

310 BUCHOLZ, W.H.S. Chymische Versuche uber das Meyerische Acidum pingue. Weimar: Bey Carl Ludolf Hoffmann, 1771. [8], 96 p.

311 JACQUIN, N.J. Chymische Untersuchung der Meyerischen Lehre von der fetten Saure, und der Blackischen von der figirten Luft. Aus dem Lateinischen ubersetzt. Leipzig: Zu finden im Krausischen Buchladen in Wien, 1771. [4], 143 p.

Fourcroy states that there were French and English translations (*Encyclopedie Methodique: Chimie, Pharmacie et Metallurgie*, tome 3, p. 373 col. 2); in fact a French precis appeared (see no. 318).

312 WEIGEL, C.E. Observationes chemicale et mineralogicae . . . Goettingae: Aere Dieterichiano, [1771] [6], 78, [4] p.; 1 plate.

W.'s criticism of Well's *Rechtfertigung* (no. 313).

313 WELL, J.J. von. Rechtfertigung der Blackischen Lehre von der figirten Luft, gegen die vom Herrn Wiegleb, Apotheker in Langensalza, darwider gemachten Einwurfe. Wien: Zu finden im Krausischen Buchladen, 1771. 8 leaves, 164 p.

314 WIEGLEB, J.C. Kleine chymische Abhandlungen von

dem grossen Nutzen der Erkenntniss des Acidi pinguis
. . . Zweyte Auflage. Langensalza: In Joh. Chr. Martini
Verlag, 1771. 182 p.

315 WELL, J.J. von. Forschung in die Ursache der Erhitzung
des ungeloschten Kalks; nebst einigen freymuthigen
Gedanken uber die dessen Erhitzung bewirken sollende
Feuermaterie. Wien: Zu finden im Krausischen
Buchladen, 1772. 64 p.
W.'s response to Weigel's *Observationes*, [1771] (no. 312).

316 Precis de la doctrine de M. Meyer, sur l'acidum pingue.
Observations sur la Physique, sur l'Histoire Naturelle et
sur les Arts, tome 2 [part. 7] (juil. 1773) 30 – 41.

317 Precis d'un ouvrage, intitule: Examen doctrinae, &c.
Examen de la doctrine de M. Meyer, touchant l'acidum
pingue; & de celle de M. Black, sur l'air fixe concernant
la chaux. Par M. Crantz . . .
Observations sur la Physique, sur l'Histoire Naturelle et
sur les Arts, tome 2 [partie 8] (aoust 1773) 123 – 139.

318 Precis raisonne du memoire de Jacquin . . . dans lequel
cet auteur discute la doctrine de M. Meyer sur l'acidum
pingue, & etablit, par une suite d'experiences, celles du
docteur Black sur l'air fixe, relativement a la chaux &
aux alkalis caustiques, pour servir a l'histoire de l'air
fixe & de l'air considere comme element des corps solides.
Observations sur la Physique, sur l'Histoire Naturelle
et sur les Arts, tome 1 [partie 2] (fev. 1773) 123 –
134.

319 FOURCY, – de. Observations sur le tableau du produit
des affinites chymiques.
Observations sur la Physique, sur l'Histoire Naturelle
et sur les Arts, tome 1 [partie 3] (mars 1773) 197 – 204.

320 FOURCY, –de. Reponse au precis raisonne du memoire

de M. Jacquin . . . en faveur de l'air fixe, contre la doctrine de M. Meyer, relativement a l'acidum pingue, insere dans le Journal de Physique du mois de fevrier 1773.
Observations sur la Physique, sur l'Histoire Naturelle et sur les Arts, tome 2 [partie 9] (sept. 1773) 218 – 245.
> German summary in *Neuesten Entdeckungen in der Chemie* 11 (1783) 182 – 205.

321 KRENGER, – . Memoire sur l'existence de l'air dans les mineraux, avec des experiences qui prouvent que quelques-uns n'ont point d'acide.
Observations sur la Physique, sur l'Histoire Naturelle et sur les Arts, tome 2 [part. 12] (dec. 1773) 466 – 473.

322 WEIGEL, C.E. Observationes chemicae et mineralogicae. Pars secunda. Gryphiae: Apud Ant. Ferd. Rose, 1773. [4], 99, 4 p.; 2 plates.
> W.'s response to Well's *Forschung* (no. 315).

323 LAVOISIER, A.L. Opuscules physiques et chymiques. 1774.
See no. 214.

324 VOGEL, R.A. Lehrsatze der Chemie. Ins deutsche ubersetzt, und mit Anmerkungen versehen von Johann Christian Wiegleb. Weimar: Bey Carl Ludolf Hoffmann, 1775.
> Section 137, Wiegleb's response to Well's and Bucholz's criticisms (nos. 313, 315, 310).

325 ERXLEBEN, J.C.P. Ueber die fixe Luft und die fette Saure: eigene Untersuchungen und Prufungen des von Andern daruber beygebrachten.
In his: Physikalisch-chemische Abhandlungen, Bd. 1, pp. 1 – 279. Leipzig: In der Weygandschen Buchhandlung, 1776.

326 LAVOISIER, A.L. Essays physical and chemical. 1776.
See no. 217.

327 [DANIEL, C.F.] Versuch einer Theorie der wichtigsten Beobachtungen aus der Naturlehre, die man zum Theil durch die fixe Luft oder fette Saure zu erklaren bemuht war. Halle, 1777. 134 p.
> D.'s response to Erxleben (nos. 325, 328).

328 ERXLEBEN, J.C.P. Experimenta nonnulla Blackianam de aere fixo doctrinam spectantia. Novi Commentarii Societatis Regiae Scientiarum Gottingensis 7 (1777) 81 —

329 WENZEL, C.F. Lehre von der Verwandschaft der Korper. Dressden: Gedruckt bey Gotthelf August Gerlach, 1777.

330 LANGMAJER, I.J. Supplementum in Ioannis Iacobi de Well defensionem doctrinae Blackianae et Epicrisin super calcis incalescentia quod praemisso utroque libello de Germanico in Latinum translato . . . edidit Ignat. Iosephus Langmajer. Vindobonae: Ex Officina Krausiana, 1778. 2 leaves, [20], 352 p.; 1 plate, facing p. 47.
> Contents: Ioannis Iacobi de Well . . . Apologia doctrinae Blackianae de aere fixo contra objectiones editas a Ioanne Christiano Wiegleb . . . supplemento aucta (pp. [5] — [24], 1 — 152); Ioan. Iac. de Well . . . Examen causae incalescentiae calcis vivae cum epicrisi de materia ignis . . . supplemento auctum (pp. 153 — 206); Ignat. Josephi Langmajer . . . Supplementum in praemissos libellos super aere fixo et causa incalescentiae calcis . . . (pp. 207 — 352).

331 MACQUER, P.J. Dictionnaire de chimie . . . , seconde edition, tome 1, pp. 290 — 333 [of the 4-vol. issue] Paris: T. Barrois, le jeune, 1778.
> "Causticite."
> This article on caustification is on pp. 192 — 219 of tome 1

of the 2-vol. issue Paris: De l'Imprimerie de Monsieur, 1778.

332 NEUFVILLE, Z. Tetamen [sic] medicum inaugurale, de natura aeris fixi, ejusque dotibus . . . Edinburgi: Apud Balfour et Smellie, 1678 [sic, i.e. 1778]. [4], 60 p.

333 WEBER, J.A. Neuentdeckte Natur und Eigenschaften des Kalkes und der azenden Korper nebst einer oconomisch- chemischen Untersuchung des Kochsalzes und dessen Mutterlauge. Berlin: A. Wever, 1778. 237 p.
"concluding that Black's and Meyer's theories are both false." − J.R. Partington in *A history of chemistry*, vol. 3, p. 152. London, 1962.

334 WEIGEL, C.E. Chemisch-mineralogische Beobachtungen . . . Aus dem Lateinischen ubersetzt und mit vielen Zusatzen vermehrt von D. Johann Theodor Pyl. Breslau: Bey Willhelm Gottlieb Korn, 1779. [6], 92, [2], 182, [2] p.; 3 plates.
Analysed in *Chemisches Journal* 3 (1780) 208 − 218.

334a SCHEELE, C.W. Uber das brennbare Wesen im rohen Kalk.
Neuesten Entdeckungen in der Chemie 1 (1781) 30 − 41.
S.'s comments on Weber's *Neuentdeckte Natur* . . . *des Kalkes* (no. 333).
English translation in DOBBIN, L., ed., *The collected papers of Carl Wilhelm Scheele*, pp. 318 − 325. London: G. Bell & Sons Ltd., 1931 (Reprinted New York: Kraus Reprint Co., 1971).

335 WENZEL, C.F. Lehre von der Verwandschaft der Korper, pp. 253 − 290. Dressden: Bey Johann Samuel Gerlach, 1782.
Section VIII, pp. 253 − 290 "Abhandlung von der fixen Luft und fetten Sauere".

336 LAVOISIER, A.L. Physikalisch-chemische Schriften. 1783.
See no. 233.

336a WEBER, J.A. Bemerkungen uber das brennbare Wesen im rohen Kalke.
Neuesten Entdeckungen in der Chemie 12 (1784) 94 – 111.
W.'s response to Scheele's criticisms (no. 334a).

336b SCHEELE, C.W. Erlauterung uber einige, den ungeloschten Kalk betreffende, Versuche.
Chemische Annalen (1785 Bd. 2) 220 – 227.
On Weber's experiments.
English translation in DOBBIN, L., ed., *The collected papers of Carl Wilhelm Scheele*, pp. 335 – 339. London: G. Bell & Sons Ltd., 1931 (Reprinted New York: Kraus Reprint Co., 1971). A previous translation had appeared in *Crell's Chemical Journal* 2 (1792) 113 – 122.

337 GUYTON DE MORVEAU, L.B. Acidum pingue, causticum.
Encyclopedie Methodique: Chymie, Pharmacie et Metallurgie 1 (1786) 418 – 420.

338 GREN, F.A.C. Systematisches Handbuch der gesammten Chemie, Theil 1, Sections 274 – 277, pp. 181 – 183.
Halle: Im Verlage der Waisenhaus-Buchhandlung, 1787.
Account of the Black/Meyer controversy.

339 JACQUIN, N.J. Chymische Untersuchung der Meyerischen Lehre von der fetten Saure . . . Wien, 1790.

340 WIEGLEB, J.C. Geschichte des Wachsthums und der Erfindungen in der Chemie, in der neuern Zeit, Bd. 2, pp. 139 – 143. Berlin und Stettin: Bey Friedrich Nicolai, 1791.

Black/Meyer Controversy

Account of the Black/Meyer controversy.

341 GREN, F.A.C. Systematisches Handbuch der gesammten Chemie, Zweite . . . Auflage, Theil 1, Sections 437, pp. 288 ff. Halle: Waisenhaus-Buchhandlung, 1794.

342 FOURCROY, A.F. de. Chimie. 1795/96.
See no. 255.

343 LAVOISIER, A.L. Opuscules physiques et chimiques. 1801.
See no. 263.

344 [THOMSON, T.] Chemistry.
Supplement to the Third Edition of the Encyclopaedia Britannica, vol. 1 (1801) pp. 212 − 403.
> In particular p. 301 for the theories of B. and Meyer.
> Thomson is identified as the author of this article in the Advertisement in vol. 2.
> 2nd edition of the *Supplement*, 1803.

345 FISCHER, J.C. Geschichte der Physik seit der Wiederherstellung der Kunste und Wissenschaften bis auf die neuesten Zeiten, Bd. 5, pp. 187 − 220. Gottingen: Bey Johann Friedrich Rower, 1804.
(Geschichte der Kunste und Wissenschaften, Abth. 8: Geschichte der Naturwissenschaften, I: Geschichte der Naturlehre, Bd. 5)
> Part of chap. 3 "Beobachtungen und Entdeckungen in der Lehre von den Gasarten".
> Account of the Black/Meyer controversy.

Later accounts of the Black/Meyer controversy are entered in Part IV of this bibliography; in particular see no. 580.

Part IV

WORKS ON BLACK

401 Address to the citizens of Edinburgh. Edinburgh, 1764. 26 p.
> Urging the transfer of Dr. Cullen to the chair of the Practice of Medicine and Dr. Black to the chair of Chemistry in Edinburgh University; B. was teaching in Glasgow at this time.

402 Address of the students of medicine, to the Right Hon. the Lord Provost, Magistrates, and Town-Council of the City of Edinburgh. [Edinburgh], 1766. 8 p.
> Urging the appointment of Dr. Gregory to the chair of the Theory of Medicine, Dr. Cullen to the chair of the Practice of Medicine, and Dr. Black to the chair of Chemistry. Signed by a committee comprising J. Maddocks, A.M. Drummond, T. Smith, and J. Blair.

403 Edinburgh Advertiser, no. 2,006 (Tues. 18th Mar. to Fri. 21st Mar. 1783) 179.
> Under the heading "Extract of another letter from London, March 17" is a notice of the appointment of W. Robertson and B. as honorary members of the new Academy of Sciences in St. Petersburg.

404 Scots Magazine 51 (Oct. 1789) 520.
> Under the heading "Preferments" is a notice of the appointment of B. as one of the six Associes Etrangers of the Academie Royale des Sciences, Paris.

405 JUNIPER, J., *pseud*. The Brunoniad; an heroic poem, in six cantos, containing a solemn detail of certain commotions which have, of late, divided the kingdom of physic against itself . . . London: Printed for G. Kearsley, 1789.
> Variously attributed to Thomas Foster, B.A., and to William Margetson Heald. A contribution to the conflict between John Brown, former Edinburgh student, and the medical

faculty; some lines referring to Black are quoted in LAWRENCE, C.L., Joseph Black: the natural philosophical background, pp. 1, 4 (in: SIMPSON, A.D.C., ed., *Joseph Black, 1728 – 1799: a commemorative symposium.* Edinburgh, 1982).

406 ANON. Some account of Dr. Joseph Black.
European Magazine, and London Review, vol. 22 [no. 2] (Aug. 1792) 83 – 84; pl. facing p. 84.

407 ANON. Some account of Dr Joseph Black.
Historical Register; or Edinburgh Monthly Intelligencer 1 (1792).

408 Caledonian Mercury, no. 12,206 (Sat. 14th Dec. 1799) 3.
This newspaper is also referred to as the *Edinburgh Mercury.* In col. 4, under the heading "Died", is a notice of B.'s death; in col. 3 the editors "inform the public, that some account of Dr. Black's studies will soon be given by a near relative . . ."

409 Edinburgh Evening Courant, no. 13,722 (Sat. 14th Dec. 1799).
Under the heading "Edinburgh – December 14" is a notice of B.'s burial in Greyfriars, on the 13th.

410 Philosophical Magazine, vol. 5 [no. 3] (Dec. 1799) 312.
Under the heading "Died" is a brief announcement of B.'s death.

411 Scots Magazine 61 (Appendix 1799) 909.
Under the heading "Deaths", at 6th Dec., is a brief account of B.'s funeral procession from Nicholson St. to Greyfriars Churchyard.

412 ANON. Black.
Allgemeines Journal der Chemie 6 (1801) 346 – 351.
Supplement to Scherer's obituary of B. (no. 416); this supplement being derived from *Annalen der Neuesten*

Brittischen Arzneykunde und Wundarzneykunst 1 (1801) 171 – 176.

413 Annalen der Neuesten Brittischen Arzneykunde und Wundarzneykunst 1 (1801) 171 – 176. See no. 412.

414 ANON. Black.
Magasin Encyclopedique, annee 7 tome 2 (1801) 394 – 396.

415 ANON. Some account of the life of the late Dr. Joseph Black.
Philosophical Magazine 10 (1801) 157 – 158.
"probably by Alexander Tilloch". – H. Guerlac in *Dictionary of Scientific Biography* 2 (1970) p. 182 col. 2.

416 SCHERER, A.N. Joseph Black.
Allgemeines Journal der Chemie 6 (1801) 98 – 112; frontispiece.
Obituary, signed "S"; identified as the editor of the journal (A.N. Scherer) on p. iii of the volume contents list. For a supplement see no. 412.

417 STEWART, D. Account of the life and writings of Thomas Reid, D.D. F.R.S.Edin. Edinburgh: Printed by Adam Neill and Company, 1802. [4], 164 p.
Brief references only, to Reid's acquaintance with B. and attendance at his lectures (pp. 34, 36).
Stewart's *Account* was reprinted in 1803; and also appears in his *Biographical memoirs*, 1811; the latter appear also as vol. 10 of his collected works; the *Account* is also prefixed to the various editions of Reid's collected works, etc.

418 ANON. Biographical account of Dr. Black.
Scots Magazine 65 (1803) 367 – 372; 439 – 445.

Works on Black

419 [BROUGHAM, H.P., *1st Baron Brougham and Vaux*.] Lectures on the elements of chemistry, delivered in the University of Edinburgh, by the late Joseph Black, M.D. . . .
Edinburgh Review 3 (1803) 1 – 26.
> A review of Black's *Lectures*; for De Luc's comments on this review see no. 422. Other reviews are listed under no. 71.

419a IRVINE, W. A letter from Mr. Irvine concerning the late Dr. Irvine, of Glasgow, his doctrine, which ascribes the disappearance of heat, without increase of temperature, to a change of capacity in bodies, and that of Dr. Black, which supposes caloric to become latent by chemical combination with bodies; with particular remarks on the mistakes of Dr. Thompson, in his accounts of these doctrines.
Journal of Natural Philosophy, Chemistry, and the Arts 6 (1803) 25 – 31.
> Commenting on T. Thompson's article on Chemistry in the *Supplement to the third edition of the Encyclopaedia Britannica*. Edinburgh, 1801 (2nd ed. 1803).

420 [ROBISON, J.] Some account of Dr. Black.
Literary Journal 1 (1803) cols. 651 – 656.
> Derived from R.'s preface to his edition of B.'s *Lectures*. Similar accounts of varying fullness appear in reviews in other literary magazines; some are listed under no. 71.

421 Bibliotheque Britannique, Sciences et Arts 28 (1805) 133 – 146, 324 – 342.

422 LUC, J.A. de. To the conductors of the Edinburgh Review.
Edinburgh Review 6 (1805) 502 – 515 (see also p. 501 for the Review editors' introductory remarks on this letter.)
> De Luc's response to Brougham's review of Black's *Lectures* (no. 419).

423 FERGUSON, A. Minutes of the life and character of Joseph Black, M.D.
Transactions of the Royal Society of Edinburgh, vol. 5 pt. 3 (1805) 101 – 117.
Read on 3rd Aug. 1801.

423a D., E. Letter of inquiry from a correspondent, respecting the spontaneous recovery of the edge in dull razors laid aside for a time; and a postscript, shewing that Lavoisier has no title to the discovery of the modern theory of oxidation: with a reply and some remarks by W.N.
Journal of Natural Philosophy, Chemistry, and the Arts 14 (1806) 89 – 94.
> Briefly mentions B.'s interest in the razor phenomenon (p. 89).

423b D., K.H. Observations and enquiries concerning the heat of air blown from bellows.
Journal of Natural Philosophy, Chemistry, and the Arts 13 (1806) 170 – 173.
> Commenting on B.'s account of this phenomenon (*Lectures on the elements of chemistry*, vol. 1, p. 88. Edinburgh, 1803).

424 M.B.L.S. A sketch of the life of Joseph Black, M.D. F.R.S. of London and Edinburgh.
Belfast Monthly Magazine 2 (1809) 442 – 448.
> The author claims personal acquaintance with B. and many branches of his family; pp. 442 – 444 contain genealogical information.

425 THOMSON, T. On the discovery of the atomic theory.
Annals of Philosophy 3 (1814) 329 – 338.
> For B. see p. 334; T. reports on B.'s use of diagrams, from his own knowledge as B.'s student and from Robison.

426 THOMSON, T. Biographical account of Joseph Black, M.D. F.R.S.E. &c. Professor of Chemistry in the

University of Edinburgh.
Annals of Philosophy 5 (1815) 321 – 327.

427 ROBISON, J. The articles steam and steam-engines, written for the Encyclopaedia Britannica, by the late John Robison . . . forming part of Dr. Robison's works, edited by David Brewster . . . with notes and additions, by James Watt, pp. iii – x. Edinburgh: Printed by James Ballantyne & Co. for John Murray . . . London, 1818.
> Prints a letter from Watt to Brewster, dated May 1814, denying Black's influence on the improvements to the steam engine; see also pp. 113 – 121, footnotes, for W.'s own account of his improvements.
>
> This work was privately issued by W.; it subsequently appeared as part of vol. 2 of ROBISON, J., *A system of mechanical philosophy* (no. 428).

428 ROBISON, J. A system of mechanical philosophy. By John Robison. With notes by David Brewster, vol. 2, pp. iii – x. Edinburgh: Printed for John Murray, London, 1822.
> Prints a letter from Watt to Brewster, dated May 1814, denying Black's influence on the improvements to the steam engine; see also pp. 113 – 121, footnotes, for W.'s own account of his improvements.
>
> Vol. 2 pp. iii – x and 1 – 184 had already been issued privately by W. under the title *The articles steam and steam engines* . . . (no. 427).

429 ANDERSON, C. Dr. Charles Anderson's machine for measuring small quantities of fluids.
Edinburgh Philosophical Journal 8 (1823) 418 – 419.
> Briefly mentions B.'s proposal for standardising the lips of apothecaries' phials (p. 419).

430 [THOMSON, T.] Black, Dr Joseph.
Edinburgh Encyclopaedia, vol. 3 (1830) pp. 548 – 553.
> This article is signed C; the author is not identified in the list

of contributors, but is given as T. Thomson in A.M. Clerke's article in the *Dictionary of National Biography* 5 (1886) 109 – 112.

431 THOMSON, T. The history of chemistry, vol. 1. London: Henry Colburn, and Richard Bentley, 1830.
> Chap. 9, pp. 303 – 349, "Of the foundation and progress of scientific chemistry in Great Britain"; in particular pp. 313 – 336 for B.

432 SINCLAIR, J. The correspondence of the Right Honourable Sir John Sinclair, Bart., vol. 1, pp. 433 – 435; vol. 2, pp. 85 – 87. London: Henry Colburn and Richard Bentley, 1831.
> S. reports his conversation with J. Montgolfier, Christmas 1785, during which M. stated that the Montgolfiers' balloon experiments were inspired by B.'s writings. Quoted by E. Cohen in *Janus* 14 (1909) 304 – 310; and in CLOW, A. & CLOW, N.L., *The chemical revolution*, pp. 153 – 154. London, 1952.

433 THOMSON, J. An account of the life, lectures and writings, of William Cullen, M.D. Vol. 1. Edinburgh: William Blackwood; London: T. Cadell, 1832. xvi, 668 p.
> Prints three letters from B. to C., 1753 – 1755 (pp. 573 – 580).
>
> No more of this issue appeared; this 1832 volume was reissued in 1859 as vol. 1 of the first complete 2-volume edition (no. 453).

434 CHAMBERS, R. A biographical dictionary of eminent Scotsmen, vol. 1, pp. 211 – 215. Glasgow: Blackie & Son [etc.], 1835.
> Vol. 1 has a volume title page *Lives of illustrious and distinguished Scotsmen* . . . vol. 1. Glasgow: Blackie & Son [etc.], 1832.
>
> Many later issues and editions.

Works on Black

435 LOCKHART, J.G. Memoirs of the life of Sir Walter Scott, Bart., vol. 1, p. 14. Edinburgh: Robert Cadell; London, John Murray and Whittaker and Co., 1837.
> Brief account of B.'s part in saving the infant Scott's life. Quoted in DOIG, A., Dr Black, a remarkable physician, p. 38 (in: SIMPSON, A.D.C., ed., *Joseph Black, 1728 – 1799. A commemorative symposium.* Edinburgh, 1982).

436 KAY, J. A series of original portraits and caricature etchings, by the late John Kay, miniature painter, Edinburgh; with biographical sketches and illustrative anecdotes, vol. 1, pp. 52 – 54, 56 – 57. Edinburgh: Hugh Paton; London: Smith, Elder & Co., 1838.
> Reissued in 1842 with improved biographies; new ed. 1877. A selection of the biographies with inferior reproductions of the plates was issued as PATERSON, J., *Kay's Edinburgh portraits.* London: Hamilton, Adams, & Co.; Glasgow: Thomas D. Morison, 1885. 2 v.; this omits the Black portraits and biography, but includes the anecdotes on Black and Hutton (vol. 1, pp. 48 – 50). See nos. 906 – 908 for details of Kay's portraits of Black.

437 K., W.A. On Dr. Black's theory of latent heat. Mechanics' Magazine 33 (1840) 21 – 23.
> The first letter in a trivial correspondence on the nature of latent heat; followed by letters from Latent *pseud.* (pp. 123 – 125) and G.A. Widney (pp. 227 – 231). They have no real reference to B.

438 ANON. Lives of men of letters and science who flourished in the time of George III. By Henry Lord Brougham . . . British Quarterly Review 2 (1845) 197 – 263.
> Review of Brougham's *Lives* (no. 439); Black is discussed on pp. 233 – 243.

439 BROUGHAM, H.P., *1st Baron Brougham and Vaux.* Lives of men of letters and science, who flourished in

Works on Black

the time of George III, pp. 324 — 351; pl. facing p. 324.
London: Charles Knight and Co., 1845.

For reviews see nos. 438, 440, 441, 443; for Brougham's supplementary note see no. 442.
Information on later editions is given under no. 442.

440 [LOCKHART, J.G.] Lives of men of letters and science who flourished in the time of George III. By Henry, Lord Brougham . . . Quarterly Review 76 (1845) 62 — 98.

A review of Brougham's *Lives* (no. 439), almost entirely of the men of letters section; B. is mentioned only once, on p. 97. The reviewer indicates that the men of science section will be discussed in a later review — see no. 441.

441 [PEACOCK, G.] 1. Eloge historique de James Watt. Par M. Arago . . . 2. Address of the Rev. William Vernon Harcourt . . . 3. Lives of men of science of the time of George III. By Henry Lord Brougham . . . Quarterly Review 77 (1845) 105 — 139.

A joint review of these three works, including the men of science section of Brougham's *Lives* (no. 439) — for a review of the men of letters section see no. 440.

442 BROUGHAM, H.P., *1st Baron Brougham and Vaux*. Lives of men of letters and science, who flourished in the time of George III, vol. 2, pp. 507 — 516. London: Henry Colburn, 1846.

"Note to the lives of Cavendish, Watt, and Black, published in the first volume" [no. 439].

The *Lives of men of letters and science* was reprinted several times in Britain, America and France, usually in 2 vols., comprising the two series. Later the chapters on the scientists were collected from both series and republished as *Lives of philosophers of the time of George III*, also constituting vol. 1 of Brougham's *Works*; this *Lives of philosophers* also went through a number of reprints and new editions. The chapter on Black is reprinted (from the 1845

Paris edition) in: FARBER, E., ed., *Great chemists*, pp. 211 – 225. New York, 1961.

443 HARCOURT, W.V. Letter to Henry Lord Brougham, F.R.S., &c., containing remarks on certain statements in his lives of Black, Watt and Cavendish. London, Edinburgh, and Dublin Philosophical Magazine and Journal of Science 3rd ser. vol. 28 (1846) 106 – 131.

444 MUIRHEAD, J.P. Correspondence of the late James Watt on his discovery of the theory of the composition of water. London: John Murray; Edinburgh: William Blackwood and Sons, 1846. cxxvii, 264 p.
> Prints extracts from the Watt – Black correspondence (9 letters, 1782 – 1784).

445 REID, T. The works of Thomas Reid . . . collected . . . by Sir William Hamilton. 1846.
> This edition not seen; see no. 455.

446 WILSON, G. A few unpublished particulars concerning the late Dr Black. Proceedings of the Royal Society of Edinburgh 2 (1849) 238.
> W. presented information on B. gathered from Mrs. E. Wordsworth; only the briefest details are given here.

447 MUIRHEAD, J.P. The origin and progress of the mechanical inventions of James Watt illustrated by his correspondence with his friends and the specifications of his patents. London: John Murray, 1854. 3 v.
> Prints "History of Mr. Watt's improvement of the steam-engine. By Joseph Black, M.D." (vol. 1, pp. xxxv – xl); 13 letters from W. to B., 1778 – 1798 (vol. 2); a letter from B. to W., 15th Mar. 1780 (vol. 2, pp. 118 – 120); and 4 letters between Robison and W. on the occasion of B.'s death, 11th – 29th Dec. 1799 (vol. 2, pp. 261 – 267).

Works on Black

448 COCKBURN, H.T., *Lord*. Memorials of his time, pp. 48 – 49. Edinburgh: A. and C. Black, 1856.
> Description of B. Quoted in: WOOD, A., *Thomas Young, natural philosopher, 1773 – 1829*, p. 29 note 3. Cambridge: University Press, 1954.

449 MUIRHEAD, J.P. The life of James Watt, with selections from his correspondence. London: John Murray, 1858. xvi, 580 p.
> Based on the author's *The origin and progress of the mechanical inventions of James Watt* (no. 447), this more popular biography does not reprint the correspondence; and reprints only part of the History of Mr. Watt's improvement of the steam-engine. By Joseph Black, M.D. (pp. 58 – 59, 95 – 96).
> 2nd ed. of the *Life*: 1859.

450 PLAYFAIR, L. A century of chemistry in the University of Edinburgh: being the introductory lecture to the course of chemistry in 1858. Edinburgh: Printed by Murray and Gibb, 1858. 32 p.
> Includes details of the transfer of apparatus associated with B. from the University to the Industrial Museum of Scotland.

451 BLOXAM, T. On the composition of old Scotch glass. By Mr Thomas Bloxam, Assistant Chemist to the Industrial Museum. Communicated, with a preliminary note, by Professor George Wilson. Proceedings of the Royal Society of Edinburgh 4 (1859) 191 – 196.
> Includes the analysis of a bottle from B.'s laboratory.

452 SCIENCE AND ART DEPARTMENT OF THE COMMITTEE OF COUNCIL ON EDUCATION. Sixth report of the Science and Art Department of the Committee of Council on Education. London: Her Majesty's Stationery Office, 1859.
> Report for the year 1858. Appendix F, pp. 101 – 106, is

Works on Black

the Annual report of the Director of the Industrial Museum of Scotland [George Wilson] ; the acquisition of laboratory relics of B. is briefly recorded on p. 103.

453 THOMSON, J. An account of the life, lectures, and writings of William Cullen, M.D. Edinburgh: W. Blackwood, 1859. 2 v.
: Vol. 1 is a reissue of the volume originally published in 1832 (no. 433).

454 WEBSTER, T. The case of Henry Cort, and his inventions in the manufacture of British iron. No. VI − VIII. Mechanics' Magazine, new ser. vol. 2 (1859) 52 − 53; 85 − 86; 100 − 102.
: p. 52 reprints an extract from Watt's letter to B., dated 6th June 1784, on C.'s process (from MUIRHEAD, J.P., *The origin and progress of the mechanical inventions of James Watt*, vol. 2, pp. 188 − 190 (see no. 447)).
: pp. 85 − 86, 100 − 102 reprint substantial extracts from *A brief state of facts relative to the new method of making bar iron*, 1787; in particular Extract of a letter from Doctor Joseph Black . . . and Extracts from Dr. Black's "Remarks . . . " are reprinted in full on pp. 101 − 102.
: p. 102 prints a letter from James Black to Lord Stanhope, dated 24th May 1786; this is the covering letter which accompanied the submission to S. of B.'s letter.

455 REID, T. The works of Thomas Reid, D.D. now fully collected, with selections from his unpublished letters. Preface, notes and supplementary dissertations, by Sir William Hamilton, Bart., sixth edition, vol. 1, pp. 41 − 48. Edinburgh: Maclachlan and Stewart; London: Longman, Green, Longman, Roberts, and Green. 1863.
: Prints seven letters from Reid to David and Andrew Skene, 1765 − 1767, referring to B.'s theories of heat and removal to Edinburgh (Letters II, III, V − VIII, X).

456 BROWN, J. The epitaphs and monumental inscriptions in

Greyfriars Churchyard, p. 219. Edinburgh: J.M. Miller, 1867.
>   Latin inscription from B.'s monument, with English translation. [The original was replaced by Edinburgh Town Council in 1894; there is a photograph of the later inscription in *Chemistry and Industry* (7 Feb. 1976) 103; also a photograph of the later monument, with an English translation, in *American Physics Teacher* 7 (1939) 126; 128. See also no. 506.]

457 BROWN, A.C. The development of the idea of chemical composition, pp. 19 – 30. Edinburgh: Edmonston and Douglas, 1869.

458 BROUGHAM, H.P., *1st Baron Brougham and Vaux*. The life and times of Henry, Lord Brougham. Written by himself, vol. 1, pp. 71 – 76. Edinburgh; London: W. Blackwood and Sons, 1871.

459 ANDREWS, T. Chemistry.
Report of the forty-first Meeting of the British Association for the Advancement of Science; held at Edinburgh in August 1871, Transactions of the Sections, pp. 59 – 66. London: John Murray, 1872.
>   Presidential address to the Chemistry Section of the Association. B. is discussed briefly on pp. 59 – 60; the three letters from Lavoisier to B. forming an appendix to this address are printed elsewhere in the 1871 *Report* (no. 460).

460 Letters from M. Lavoisier to Dr. Black.
Report of the forty-first Meeting of the British Association for the Advancement of Science; held at Edinburgh in August 1871, Reports on the State of Science, pp. 189 – 192. London: John Murray, 1872.
>   Three letters, dated 19th Sept. 1789, 24th July 1790 and 19th Nov. 1790; these letters form an appendix to T. Andrews' address which is printed elsewhere in the *Report*

(no. 459).
For later republications see nos. 461, 468, 492, 526.

461 Letters from M. Lavoisier to Dr. Black. [s.l.:s.n., 1872?] 3 p.
>A separate issue of no. 460; in the same typeface as the latter, but with re-set title, omission of the running title, and re-structured pagination.
Copy in the Edgar Fahs Smith Memorial Collection, University of Pennsylvania Library, in THORPE, T.E., compiler, Miscellaneous chemical memoirs, vol. 4.

462 PLAYFAIR, L. Air and airs, as illustrated by the Magdeburg hemispheres and Black's and Cavendish's balances.
In: SOUTH KENSINGTON MUSEUM. Free evening lectures, delivered in connection with the Special Loan Collection of Scientific Apparatus, 1876, pp. 134 – 154. London: Chapman and Hall, [1876].
>Lecture delivered 3rd July 1876; B. is discussed on pp. 144 – 147.

463 SCIENCE AND ART DEPARTMENT OF THE COMMITTEE OF COUNCIL ON EDUCATION. Catalogue of the Special Loan Collection of Scientific Apparatus at the South Kensington Museum, 1876, p. 476. London: Her Majesty's Stationery Office, 1876.
>No. 2401, Balance used in his experiments by Dr. Joseph Black (p. 476); no. 2537, Pneumatic trough used in his experiments by Dr. Joseph Black (p. 492). There are other English, French and German issues of this catalogue.

464 BROWN, A.C. Joseph Black.
Nature 18 (1878) 346 – 347.
>Abstract of a lecture to the Edinburgh University Chemical Society.

465 GRANT, J. Cassell's old and new Edinburgh. London:

Works on Black

Cassell & Company, Limited, [1884 – 1887]. 3 v.
> Published in parts; various issues. References to B. and places associated with him include: Smellie's printing office (vol. 1 p. 236), the College Wynd (vol. 2 pp. 254 – 255), the Royal Infirmary (vol. 2 p. 298), the Royal Blind Asylum and School (vol. 2 pp. 335 – 336; woodcut on p. 340, after Storer).

466 MONCREIFF, J., *1st Baron*. . . . Review of the hundred years' history of the Society.
Proceedings of the Royal Society of Edinburgh 12 (1884) 451 – 474.
> Brief account of B. on pp. 462 – 463.

467 CLERKE, A.M. Black, Joseph, M.D. (1728 – 1799).
Dictionary of National Biography, vol. 5 (1886) pp. 109 – 112.

468 [RICHET, C.] Experiences inedites de Lavoisier sur la respiration.
Revue Scientifique 39 (1887) 193 – 194.
> This "tome 39 de la collection" is also numbered 3eme ser. tome 13, and 24me annee 1er semestre.
> Prints a letter from Lavoisier to B., dated 13th Nov. 1790.
> This letter had already been published in no. 460 (dated 19th Nov. 1790); for later republications see nos. 492, 526.

469 RICHARDSON, B.W. Joseph Black, M.D., and the school of chemical medicine.
Asclepiad 7 (1890) 254 – 282.
> Reprinted as no. 474.

470 HARRISON, W. Memorable Edinburgh houses,
pp. 29 – 30. Edinburgh: Oliphant Anderson and Ferrier, 1893.
> No. 58 Nicolson Street, where B. had lived; by 1893 it had become the Asylum for the Blind. A very brief description.
> Revised ed., 1898; reprinted 1971.

471 FINLAYSON, J. "Members of the medical profession in Glasgow of wide celebrity". Memoranda. [s.l., 1896].
Not seen; reference from *Bibliotheca Osleriana*, item 6653, p. 572. Oxford: Clarendon Press, 1929.

472 RAMSAY, W. The gases of the atmosphere: the history of their discovery, chap. 2, pp. 38 – 67. London: Macmillan and Co., Ltd., 1896.
" 'Fixed air' and 'mephitic air' – their discovery by Black and by Rutherford".
2nd ed., 1902; 3rd, 1905; 4th, 1915.

473 DOBBIN, L. Dr. Joseph Black: a centenary memorial sketch.
Chemist and Druggist 55 (1899) 942 – 944.
Reprinted as no. 523.

473a GRAHAM, H.G. The social life of Scotland in the eighteenth century, vol. 2, p. 195, note. London: Adam and Charles Black, 1899.
Prints a letter to Adam Smith, dated 1764, describing "the Reformation that has been made in our devotions since you left us." The *Social life* has been reissued often; the letter is on p. 461 of the one-vol. issues.

474 RICHARDSON, B.W. Joseph Black, M.D., and the school of chemical medicine.
In: RICHARDSON, B.W., Disciples of Aesculapius, vol. 2, pp. 461 – 481. London: Hutchinson & Co., 1900.
Reprinted from no. 469.

475 FOSTER, M. Lectures on the history of physiology during the sixteenth, seventeenth and eighteenth centuries, lecture 9, pp. 224 – 254. Cambridge: University Press, 1901.
"The rise of the modern doctrines of respiration. Black, Priestley, Lavoisier". In particular pp. 232 – 237 for B.

2nd ed., 1924.

476 INTERNATIONAL EXHIBITION GLASGOW 1901. Official catalogue of the Scottish History and Archaeology Section, p. 86. Glasgow: Chas. P. Watson, Publisher, [1901].
> No. 770, Memorials of Joseph Black, M.D. [Scientific apparatus].

477 STIRLING, W. Some apostles of physiology, pp. 65 – 67; pl. facing p. 66. London: Privately printed . . . by Waterlow and Sons Limited, 1902.

478 WARD, I.W. The Black family. Ulster Journal of Archaeology (1902) 176 – 188.
> Genealogical and other information on B.'s family – in particular on his grandfather, his father, and the descendents of his brother George.

479 RAMSAY, W. Joseph Black, M.D. A discourse . . . Delivered in the University of Glasgow on Commemoration Day, 19th April, 1904. Glasgow: James MacLehose and Sons, 1904. 26 p.; 5 plates.
> The Commemoration Day celebrations, including R.'s discourse, are briefly reported in *Nature* 69 (1904) 612 – 613; and in the *British Medical Journal* (1904 vol. 1) 974 – 975.
>
> R.'s discourse is reprinted in nos. 480 and 482.

480 RAMSAY, W. Joseph Black: his life and work. In: RAMSAY, W., Essays biographical and chemical, pp. 67 – 87. London: Archibald Constable & Co. Ltd., 1908.
> Reprinted from no. 479.
>
> R.'s *Essays* appeared also in an American edition New York: E.P. Dutton and Company, 1909; and in a German translation by W. Ostwald entitled *Vergangenes und Kunstiges aus der Chemie*. Leipzig: Akademische

Verlagsgesellschaft, 1909.

481 COHEN, E. Zur Geschichte der Erfindung des Luftballons.
Janus 14 (1909) 304 – 310.

482 RAMSAY, W. Joseph Black, M.D. A discourse.
In: DIERGART, P., ed., Beitrage aus der Geschichte der Chemie dem Gedachtnis von Georg W.A. Kahlbaum . . . gewidmet . . . , pp. 431 – 450. Leipzig: Franz Deuticke, 1909.
Reprinted from no. 479.

483 COCKBURN, H.A. An account of the Friday Club written by Lord Cockburn, together with notes on certain other social clubs in Edinburgh.
Book of the Old Edinburgh Club 3 (1910) 105 – 178.
pp. 145 – 154 "The Poker Club"; B. appears in the membership lists for 1768, 1776, and ca. 1786/1787.

484 GAHN, H. [Letters to Linnaeus from London and Edinburgh, 1771 – 1773].
In: FRIES, T.M., ed., Bref och skrifvelser af och till Carl von Linne, afd. 1, del 6, letters nos. 1371 – 1376, pp. 177 – 201. Stockholm: Aktiebolaget Ljus, 1912.
G. briefly describes B.'s current researches in heat, beginning: "Chemiae Pr. Black . . . har nu begynt med ett annat systeme om Hetta och kold . . . " (letter no. 1372, p. 187); B. is also mentioned in letter 1373 (p. 191) and 1376 (p. 199).

485 ADDISON, W.I. The matriculation albums of the University of Glasgow from 1728 to 1858, p. 34.
Glasgow: James Maclehose and Sons, 1913.
Recording B.'s matriculation on Frid. 14th Nov. 1746.

486 RAMSAY, W. The life and letters of Joseph Black, M.D. By Sir William Ramsay K.C.B., F.R.S. With an

introduction dealing with the life and work of Sir
William Ramsay by F.G. Donnan, F.R.S. London:
Constable and Company Ltd., 1918. xix, 148 p.
> Reviewed in the *Times Literary Supplement* (14th Nov.
> 1918) p. 549; and in *Nature* 103 (1919) 181.

487 COHEN, E. Chemisch-historische aanteekeningen VI.
Uit het leven van Joseph Black.
Chemisch Weekblad 16 (1919) 168 – 178.

488 JORISSEN, W.P. Joseph Black, Petrus Driessen und andere uber magnesia alba.
Chemisch Weekblad 16 (1919) 1579 – 1589.

489 RIDDELL, H. The great chemist, Joseph Black, his Belfast friends and family connections.
Proceedings of the Belfast Natural History and Philosophical Society 3 (1920) 49 – 88.

490 THORPE, T.E. Joseph Black and Belfast.
Nature 106 (1920) 165.
> Genealogical information extracted from no. 489.

491 BRITISH ASSOCIATION FOR THE ADVANCEMENT OF SCIENCE. Edinburgh's place in scientific progress. Prepared for the Edinburgh meeting of the British Association by the local editorial committee. Edinburgh; London: W. & R. Chambers, Limited, 1921.
> Chap. 2, pp. 44 – 62, "Chemistry" includes on pp. 44 – 54
> a section on "Pure Chemistry" by L.D. [Leonard Dobbin];
> B. is briefly discussed on pp. 45 – 48 and pl. facing p. 48.

492 RIDDELL, H. Dr. Thomas Andrews: the great chemist and physicist.
Proceedings of the Belfast Natural History and Philosophical Society (session 1920/1921, published 1921) 107 – 138.
> Reprints on pp. 135 – 138 the three letters from Lavoisier

Works on Black

to B. dated 1789 – 1790 originally published by Andrews as no. 460.

493 MASSON, I. Three centuries of chemistry: phases in the growth of a science, chap. 8, pp. 109 – 116. London: Ernest Benn Limited, 1925.
"The use of weight: Joseph Black, 1754."

494 BARCLAY, A. Catalogue of the collections in the Science Museum, South Kensington, with descriptive notes & illustrations. Chemistry, p. 15; pl. 1. London: His Majesty's Stationery Office, 1927.
No. 5, Replica of Black's balance.

495 EBSTEIN, E. Der medizinische Chemiker Joseph Black. Zeitschrift fur Medizinische Chemie 10/12 (1927) 119.

496 KNICKERBOCKER, W.S. Classics of modern science (Copernicus to Pasteur), chap. 12, pp. 89 – 95. New York: Alfred A. Knopf, 1927.
"Joseph Black, 1728 – 1799"; a very brief biographical note followed by a reprint of two passages from pt. 1 of *Experiments upon magnesia alba, quick-lime, and other alcaline substances.*

497 ANON. The bicentenary of Joseph Black. Nature 122 (1928) 59 – 60
By Edgar C. Smith?

498 SPETER, M. Joseph Black. Zur Erinnerung an sein Geburtsjahr 1728.
Chemiker-Zeitung 52 (1928) 913.

499 CRANSTON, J.A. Bicentenary address on Joseph Black. Proceedings of the Royal Philosophical Society of Glasgow 57 (1929) 70 – 84.

500 SPETER, M. Black.

In: BUGGE, G., ed., Das Buch der grossen Chemiker, Bd. 1, pp. 240 – 252; pl. 18. Berlin: Verlag Chemie GmbH, 1929.

501 MELDRUM, A.N. The eighteenth century revolution in science – the first phase. London: Longmans, Green and Co. Ltd., 1930. [8], 60 p.
> In particular chap. 3, pp. 14 – 34, "The Opuscules Physiques et Chimiques" [of Lavoisier].
> Reprinted as no. 628.

502 SPETER, M. Joseph Blacks "Mikrowaage" mit Reiterversatz.
Zeitschrift fur Instrumentenkunde 50 (1930) 204 – 206.

503 NEWELL, L.C. Caricatures of chemists as contributions to the history of chemistry.
Journal of Chemical Education 8 (1931) 2138 – 2155.
> Two of Kay's caricatures of B. are reproduced and briefly discussed on pp. 2138 – 2140. See nos. 906 – 908 for details of Kay's portraits of B.

504 BELL, J. Joseph Black, M.D.: his connection with Dublin.
Irish Journal of Medical Science, ser. 6 vol. 79 (1932) 370 – 374.
> "In the minutes of the Medico-Philosophical Society of Dublin, there is recorded that on Thursday, 13th October, 1767, Mr. [George] Cleghorn read a paper, communicated by Dr. Joseph Black, on a method for preparing ammoniacal copper . . . This paper was never printed . . . the above entry is the only record of its existence".

505 HUME, D. The letters of David Hume. Edited by J.Y.T. Greig, vol. 2. Oxford: The Clarendon Press, 1932.
> pp. 314 – 336 contain a number of references to B., H.'s physician during his final illness; p. 449 prints a letter from B. to Adam Smith, dated 26th Aug. 1776, describing H.'s

death; this letter had been previously printed, "with considerable alterations", as an addendum to H.'s *My own life*.

506 OESPER, R.E. Memorial tablet to Joseph Black.
Journal of Chemical Education 10 (1933) 716.
> Includes photograph, Latin text and English translation of the inscription on B.'s monument in Greyfriars churchyard, as replaced in 1894. See also no. 456.

507 DAVIS, T.L. Joseph Black (1728 – 1799).
Journal of Chemical Education 11 (1934) 485.
> Frontispiece to the Sept. 1934 issue of the *Journal*. "Contributed by Tenney L. Davis", it comprises a reproduction of the anonymous portrait in the Scottish National Portrait Gallery, with a brief caption. See no. 905 for details of this portrait.

508 KENDALL, J. The first chemical society in the world.
Journal of Chemical Education 12 (1935) 565 – 566.

509 MACKENZIE, J.E. The chair of chemistry in the University of Edinburgh in the XVIIIth and XIXth centuries.
Journal of Chemical Education 12 (1935) 503 – 511.
> B. is briefly discussed on pp. 506 – 507; his chair is illustrated on p. 506.

510 MAGIE, W.F. A source book in physics, pp. 134 – 145. New York; London: McGraw-Hill Book Company, Inc., 1935.
> "Black"; a very brief biographical note followed by reprints of three passages from *Lectures on the elements of chemistry*, London, 1803 (Specific heat, Latent heat, Of vapour and vaporisation).

511 McKIE, D. & HEATHCOTE, N.H. de V. The discovery of specific and latent heats. London: Edward Arnold

& Co., 1935. 155 p.
> In particular chap. 1, pp. 11 – 30, "The work of Joseph Black (1728 – 1799)"; and chap. 2, pp. 31 – 53, "Date and priority of Black's experiments."
> Reprinted New York: Arno Press, 1975.

512 McKIE, D. On Thos. Cochrane's MS. notes of Black's chemical lectures, 1767 – 8.
Annals of Science 1 (1936) 101 – 110; pl. XI.
> Subsequently considered to constitute the first part of McKie's series of papers entitled "On some MS. copies of Black's chemical lectures"; for the later parts see nos. 570, 572, 579, 585, 589.

513 NEAVE, E.W.J. Joseph Black, M.D. (1728 – 1799).
School Science Review 18 (1936) 45 – 49; pl. facing p. 45.

514 NEAVE, E.W.J. Joseph Black's Lectures on the elements of chemistry.
Isis 25 (1936) 372 – 390; pl. 2

515 WEEKS, M.E. Some scientific friends of Sir Walter Scott.
Journal of Chemical Education 13 (1936) 503 – 507.
> B. is discussed briefly on pp. 503 – 504.

516 KENDALL, J. Old chemical societies in Scotland.
Chemistry & Industry 15 (1937) 141 – 142 [Part of the Journal of the Society of Chemical Industry, vol. 56].

517 PARTINGTON, J.R. & McKIE, D. Historical studies on the phlogiston theory. – I. The levity of phlogiston.
Annals of Science 2 (1937) 361 – 404.
> In particular pp. 380 – 387, "The absolute levity of phlogiston."
> Reprinted as no. 629.

518 NIERENSTEIN, M. Black's simplification of Bergman's

chemical symbols.
Isis 28 (1938) 463 – 464.
Draws attention to Crell's paper in *Chemische Annalen* (1785 Bd. 1) 346 – 349. [Nierenstein's reference to 1795 is a misprint].

519 KENDALL, J. Young chemists and great discoveries.
London: G. Bell & Sons Ltd., 1939.

520 WATSON, E.C. Reproductions of prints, drawings and paintings of interest in the history of physics. 5. Portraits and caricatures of Joseph Black, and prints of Edinburgh and Glasgow in his day.
American Physics Teacher 7 (1939) 123 – 129.
Reproduces Kay's caricatures, Heath's engraving, and the anonymous crayon drawing. See nos. 905 – 908, 911 for details of these portraits.

521 SCHAEFER, G. & NAUMANN, W. Physicians as inventors of steam engines.
Ciba Symposia 3 (1941) 1033 – 1036.

522 CLOW, A. & CLOW, N.L. Lord Dundonald.
Economic History Review 12 (1942) 47 – 58.
Prints on pp. 50 – 52 a letter from B. to Andrew Stuart, dated 25th Jan. 1783, giving his opinion on Dundonald's manufacture of tar.

523 DOBBIN, L. Dr. Joseph Black: a centenary memorial sketch.
In: DOBBIN, L., Occasional fragments of chemical history, pp. 1 – 8. Edinburgh: Printed for private circulation, 1942.
Reprinted from no. 473.

524 KENDALL, J. Some eighteenth-century chemical societies.
Endeavour 1 (1942) 106 – 109.

525 KENT, A. Joseph Black, M.D. (1728 — 1799).
Chemistry & Industry 20 (1942) 530 — 531. [Part of the
Journal of the Society of Chemical Industry, vol. 61].

526 MIELI, A. Una lettera di A.L. Lavoisier a J. Black.
Archeion 25 (1943) 237 — 239.
> Vol. 25 of Archeion is also numbered Nueva serie T.4.
> The letter is reprinted, with added comments by Mieli, from no. 468.

527 MOULTON, F.R. & SCHIFFERES, J.J. The
autobiography of science, pp. 218 — 222. Garden City,
N.Y.: Doubleday, Doran and Company, Inc., 1945.
> "Joseph Black (1728 — 99)"; a brief biographical note followed by reprints of two passages, from *Lectures on the elements of chemistry*, London, 1803 (Latent heat: crux of the steam engine) and RAMSAY, W., *Essays biographical and chemical*. London, 1908 (Fixed air: carbon dioxide).

528 BARNETT, M.K. The development of the concept of
heat: from the fire principle of Heraclitus through the
caloric theory of Joseph Black. I.
Scientific Monthly 62 (1946) 165 — 172.

529 BARNETT, M.K. The development of the concept of
heat: from the fire principle of Heraclitus through the
caloric theory of Joseph Black. II.
Scientific Monthly 62 (1946) 247 — 257.

530 KENDALL, J. The first chemical journal.
Nature 159 (1947) 867.
> The MS Dissertations read before the "Chemical Society instituted in the beginning of the year 1785".

531 READ, J. Humour and humanism in chemistry, chap. 8,
pp. 158 — 176. London: G. Bell and Sons Ltd., 1947.
> Chap. 8 "Chemistry grows brighter"; pp. 158 — 176 are on B.; in particular the St. Andrews University MS lecture notes

are described. (The rest of the chapter, pp. 176 – 191, is on J. Marcet)

532 DUVEEN, D.I. Bibliotheca alchemica et chemica. London: E. Weil, 1949.
p. 81 contains a brief description of *Directions for preparing aerated medicinal waters* (no. 239); p. 533 contains a brief description of B.'s copy of *The chemical essays of Charles-William Scheele*, 1786, profusely annotated by B.; pl. 13 illustrates two pages of this copy of Scheele.

533 McKIE, D. Antoine Laurent Lavoisier, F.R.S. 1743 – 1794.
Notes and Records of the Royal Society of London 7 (1949) 1 – 41; pls. 1 – 2.
pp. 9 – 13 include the text of two letters from B. to Lavoisier, dated 24th Oct. 1790 and 28th Dec. 1790; they are discussed on p. 36. The first of these is the letter which, translated into French, appeared in *Annales de Chimie* 8 (1791) 225 – 229 (pp. 39 – 41 of McKie's paper contain a reprint of this French version).

534 CRANSTON, J.A. Black's influence on chemistry.
In: KENT, A., ed., An eighteenth century lectureship in chemistry, pp. 99 – 106. Glasgow: Jackson, Son & Company, 1950.

535 FLECK, A. The industrial development of Scotland in the Cullen–Black period.
In: KENT, A., ed., An eighteenth century lectureship in chemistry, pp. 107 – 125; pl. 9. Glasgow: Jackson, Son & Company, 1950.
In particular pp. 115 – 122, "The contributions to industry of Joseph Black".

536 KENT, A. An eighteenth century lectureship in chemistry. Essays and bicentenary addresses relating to the Chemistry Department (1747) of Glasgow University

(1451). Edited by Andrew Kent. Glasgow: Jackson, Son & Company, 1950. xv, 233 p. (Glasgow University publications, 82)
>The four contributions directly relating to B. are separately listed in the present bibliography.

537 NEVILLE, S. The diary of Sylas Neville, 1767 – 1788. Edited by Basil Cozens-Hardy. London: Oxford University Press, 1950. xvi, 357 p.
>N. attended B.'s lectures as a medical student; this edition of the diary has 12 brief references to B., 1771 – 1776. (The editor (p. xv) indicates that he has "cut down the very full account of his five years as an Edinburgh medical student".)

538 PATTERSON, T.S. Some observations and an anecdote. In: KENT, A., ed., An eighteenth century lectureship in chemistry, pp. 191 – 197. Glasgow: Jackson, Son & Company, 1950.
>The observations are on teaching methods, with some reference to B.; the anecdote does not relate to B.

539 READ, J. Joseph Black, M.D.: the teacher and the man. In: KENT, A., ed., An eighteenth century lectureship in chemistry, pp. 78 – 98; pl. 8. Glasgow: Jackson, Son & Company, 1950.

539a ROLLER, D. The early development of the concepts of temperature and heat: the rise and decline of the caloric theory, section 2, pp. 17 – 47, 94 – 97. Cambridge: Harvard University Press, 1950. (Harvard case histories in experimental science, case 3)
>"Joseph Black's discoveries of specific and latent heat." The case histories were also published in collected form as CONANT, J.B., ed., Harvard case histories in experimental science. Cambridge: Harvard, 1957.

540 COWEN, D.L. The Edinburgh dispensatories.

Papers of the Bibliographical Society of America 45 (1951) 85 – 96.
See also no. 560.

541 NEAVE, E.W.J. Chemistry in Rozier's Journal. – IV and V.
Annals of Science 7 (1951) 284 – 299.
In particular V, pp. 293 – 299; pl. 27, "Fixed air".

542 RUSH, B. Letters. Edited by L.H. Butterfield. Princeton: Published for the American Philosophical Society by Princeton University Press, 1951. 2 v. (Memoirs of the American Philosophical Society, vol. 30 pts. 1 – 2).
A few references to B., mostly in vol. 1; in particular R.'s opinion of B. and his lectures (letter to John Morgan, dated 27th July 1768, in vol. 1, pp. 61 – 62).

543 CLOW, A. & CLOW, N.L. The chemical revolution: a contribution to social technology. London: The Batchworth Press, 1952. xvi, 680 p.
Many references to B.; in particular reprints the letter from B. to Andrew Stuart, dated 25th Jan. 1783 (pp. 404 – 407); this letter had already been published in no. 522.

544 FLEMING, D. Latent heat and the invention of the Watt engine.
Isis 43 (1952) 3 – 5.

545 KENDALL, J. The first chemical society, the first chemical journal, and the chemical revolution. Proceedings of the Royal Society of Edinburgh 63A (1952) 346 – 358; pl. 1 – 2.

546 KENDALL, J. The first chemical society, the first chemical journal, and the chemical revolution (Part II). Proceedings of the Royal Society of Edinburgh 63A (1952) 385 – 400.

Includes a list of the 32 dissertations read to the Chemical Society of the University of Edinburgh, 1785 – 1786.

547 McELROY, D.D. The literary clubs and societies of eighteenth century Scotland, and their influence on the literary productions of the period 1700 to 1800. (Ph.D. thesis, University of Edinburgh, 1952).

548 FRACKELTON, W.G. Joseph Black and some aspects of medicine in the eighteenth century.
Ulster Medical Journal 22 (1953) 87 – 89.

549 KENDALL, J. Great discoveries by young chemists.
London: Thomas Nelson and Sons Ltd, 1953.
> Chap. 8, pp. 181 – 194, "The first chemical society and the first chemical journal".
> This work is a new edition of *Young chemists and great discoveries*. London: G. Bell & Sons Ltd., 1939; the 1953 edition was reprinted in 1954.

550 DOPSON, L. Bicentenary of "fixed air": Joseph Black and carbon dioxide.
Chemist and Druggist 161 (1954) 605.

551 WATSON, E.C. Reproductions of prints, drawings, and paintings of interest in the history of physics. 57. A contemporary portrait of Joseph Black.
American Journal of Physics 22 (1954) 32.
> A wax model for a Tassie medallion. See no. 922 for further details.

552 ANON. Joseph Black.
Endeavour 14 (1955) 115 – 116.
> Unsigned editorial – T.I. Williams was editor of *Endeavour* at this time.

553 BISHOP, T.H. The foundation of quantitative analysis: a consequence of the discovery of carbon dioxide by

Joseph Black, on June 5, 1755.
Chemist and Druggist 163 (1955) 621.

554 EDELSTEIN, S.M. Historical notes on the wet-processing industry. IX — Two Scottish physicians and the bleaching industry: the contributions of Home and Black.
American Dyestuff Reporter 44 (1955) 681 – 684.
Reprinted as no. 557.

555 BUESS, H. Joseph Black (1728 – 1799) und die Anfange chemischer Experimentalforschung in Biologie und Medizin.
Gesnerus 13 (1956) 165 – 189.

556 BUESS, H. Der qualitative und quantitative Nachweis der "fixen Luft" (1756), ein Wendepunkt in der Geschichte der Chemie und Biologie.
Experientia 12 (1956) 445 – 446.

557 EDELSTEIN, S.M. Two Scottish physicians and the bleaching industry: the contributions of Home and Black. In: EDELSTEIN, S.M., Historical notes on the wet-processing industry. Published in commemoration of the Perkin Centennial, 1956, pp. 35 – 38. s.l.: American Dyestuff Reporter, [1956?].
Reprinted from no. 554.

558 MILES, W.D. Joseph Black, Benjamin Rush and the teaching of chemistry at the University of Pennsylvania.
Library Chronicle 22 (1956) 9 – 18.

559 TODHUNTER, E.N. Biographical notes from the history of nutrition: Joseph Black — April 16, 1728 — November 10 [sic], 1799.
Journal of the American Dietetic Association 32 (1956) 311.

560 COWEN, D.L. The Edinburgh Pharmacopoeia.

Medical History 1 (1957) 123 — 139; 340 — 351.
> The second part is a bibliography of editions, reprints and translations of the *Pharmacopoeia*; the translations section is updated in COWEN, D.L., *The spread and influence of British pharmacopoeial and related literature*, pp. 33 — 35. Stuttgart: Wissenschaftliche Verlagsgesellschaft, 1974. (Veroffentlichungen der Internationalen Gesellschaft fur Geschichte der Pharmazie, n.F. Bd. 41). Cowen has also compiled *Library holdings of the Edinburgh pharmacopoeia*. New Brunswick: Rutgers University, Department of History and Political Science, 1957. 14 l. See also no. 540.

561 FOREGGER, R. Joseph Black and the identification of carbon dioxide.
Anesthesiology 18 (1957) 257 — 264.

562 GUERLAC, H. Joseph Black and fixed air: a bicentenary retrospective, with some new or little known material.
Isis 48 (1957) 124 — 151; pl. facing p. 132.
> Prints on pp. 149 — 150 a letter from B. to his father, dated 8th Mar. 1754.

563 GUERLAC, H. Joseph Black and fixed air. Part II.
Isis 48 (1957) 433 — 456.

564 MAGELLAN, J.H. de. No 186. Magalhaens a X . . .
In: Oeuvres de Lavoisier, Correspondance, fasc. 2, pp. 356 — 366. Paris: Editions Albin Michel, 1957.
> Letter, "probably to Trudaine de Montigny", dated 5th July 1772. "interesting report . . . on the work of the British scientists with fixed air and artificial mineral waters; Priestley, Pringle, Macbride, Black, and Lane are singled out . . . " — D.I. Duveen in *Supplement to a bibliography of the works of Antoine Laurent Lavoisier 1743 — 1794*, p. 88. London: Dawsons of Pall Mall, 1965.

565 McKIE, D. The "Observations" of the Abbe Francois Rozier (1734 — 93) — I.

Annals of Science, vol. 13 no. 2 (for June 1957, published Sept. 1958) 73 – 89.
> Reprints in full (on pp. 86 – 89) *Experiences du Docteur Black, sur la marche de la chaleur dans certain* [sic] *circonstances*, from *Observations sur la Physique, sur l'Histoire Naturelle et sur les Arts* (sept. 1772) 156 – 166 (no. 211).

566 COLE, W.A. Joseph Black's chemical lectures. (Query no. 156).
Isis 49 (1958) 439.
> For the result of this query on the location of MS copies of B.'s lectures see no. 634.

567 McKIE, D. Discovery of 'fixed air': Joseph Black's dissertation. (Great scientific papers, 4)
Times Educational Supplement (16th May 1958) 794.

568 McKIE, D. & HEATHCOTE, N.H. de V. William Cleghorn's De igne (1799).
Annals of Science, vol. 14 no. 1 (for Mar. 1958, published May 1960) 1 – 82.
> Includes a reprint of the Latin text and an English translation.

569 CROSLAND, M.P. The use of diagrams as chemical 'equations' in the lecture notes of William Cullen and Joseph Black.
Annals of Science, vol. 15 no. 2 (for June 1959, published Aug. 1961) 75 – 90.

570 McKIE, D. On some MS. copies of Black's chemical lectures – II.
Annals of Science, vol. 15 no. 2 (for June 1959, published Aug. 1961) 65 – 73.
> Includes the text of the first lecture of B.'s course at Edinburgh, 1785 – 86, from G. Cayley's notes; this was B.'s introductory lecture, not published by Robison.

571 WALLACE, J. The story of two portraits: William Cullen — Joseph Black.
Royal Scottish Society of Arts Bulletin, no. 8 (June 1959) 1 — 8.
> Presidential address, 13th Oct. 1958. Also published as no. 575.

572 McKIE, D. On some MS. copies of Black's chemical lectures — III.
Annals of Science, vol. 16 no. 1 (for Mar. 1960, published June 1962) 1 — 9.
> Includes the text of the second and third lectures of B.'s course at Edinburgh, 1785 — 86, from G. Cayley's notes; these lectures, on the history of chemistry, were for the most part not published by Robison.

573 McKIE, D. & KENNEDY, D. On some letters of Joseph Black and others.
Annals of Science, vol. 16 no. 3 (for Sept. 1960, published Nov. 1962) 129 — 170.
> Includes the texts of 21 letters from B. to his brother Alexander, ca. 1782 — 12 Jan. 1793, and five others.

574 PARTINGTON, J.R. Joseph Black's "Lectures on the elements of chemistry".
Chymia 6 (1960) 27 — 67.

575 WALLACE, J. William Cullen and Joseph Black: the story of two portraits.
Res Medica 17 (1 — 2) (1960) 33 — 40.
> Also published as no. 571.

576 FARBER, E. Great chemists, pp. 211 — 225. New York: Interscience Publishers, 1961.
> Reprint of Brougham's biography of Black; for more details see nos. 439, 442.

577 CROSLAND, M.P. Historical studies in the language of

Works on Black

chemistry. London: Heinemann, 1962. xvii, 406 p.

578 CROWTHER, J.G. Scientists of the Industrial Revolution, pp. 7 — 92. London: The Cresset Press, 1962.
> "with new insights and some inaccuracies." — H. Guerlac in *Dictionary of Scientific Biography*, vol. 2 (1970) p. 183 col. 1.

579 McKIE, D. On some MS. copies of Black's chemical lectures — IV. Annals of Science, vol. 18 no. 2 (for June 1962, published Apr. 1964) 87 — 97.
> Includes the text of the fourth and fifth lectures of B.'s course at Edinburgh, 1785 — 86, from G. Cayley's notes.

580 PARTINGTON, J.R. A history of chemistry, vol. 3, chap. 4, pp. 109 — 158. London: Macmillan & Co. Ltd; New York, St Martin's Press, 1962.
> "Hales and Black"; in particular pp. 130 — 143 for B.'s chemical work; pp. 143 — 153 for the Meyer/Black controversy etc.; pp. 153 — 156 for B.'s researches on heat.

581 PEREIRA PINTO, G. Quadros da historia da quimica. III. O honoravel Professor Black, M.D. Anais da Faculdade de Medicina da Universidade do Recife 5 (1962) 27 — 32.

582 BOSWELL, J. The ominous years, 1774 — 1776. Edited by Charles Ryskamp and Frederick A. Pottle, pp. 160, 201. London: Heinemann, 1963. (Yale editions of the private papers of James Boswell)
> References to Black's lectures (Boswell's entries for Summer Session 1775, 18th Dec. 1775, and 22nd Dec. 1775).

583 EYLES, V.A. The evolution of a chemist: Sir James Hall, Bt., F.R.S., P.R.S.E., of Dunglass, Haddingtonshire, (1761 — 1832), and his relations with Joseph Black,

Antoine Lavoisier, and other scientists of the period.
Annals of Science, vol. 19 no. 3 (for Sept. 1963, published Mar. 1965) 153 – 182; pl. 11.

584 LEICESTER, H.M. & KLICKSTEIN, H.S. A source book in chemistry 1400 – 1900, pp. 80 – 91. Cambridge, Massachusetts: Harvard University Press, 1965.
> "Joseph Black (1728 – 1799)"; a brief biographical note followed by reprints of passages from *Experiments on magnesia alba, quicklime, and some other alcaline substances.*

585 McKIE, D. On some MS. copies of Black's chemical lectures – V.
Annals of Science, vol. 21 no. 4 (for Dec. 1965, published July 1966) 209 – 255.
> A comparison of G. Cayley's MS notes, University College manuscript MS Add. 96, and Robison's published text.

586 ANON. Joseph Black – rediscoverer of fixed air.
Journal of the American Medical Association 196 (1966) 362 – 363.

587 ANON. Joseph Black papers for University.
University of Edinburgh Journal 23 (1967/1968) 116 – 117.

588 DYCK, D.R. The nature of heat and its relationship to chemistry in the eighteenth century. (Ph.D. thesis, University of Wisconsin, 1967)
> In particular chap. 5, pp. 125 – 172, "Joseph Black and the concepts of latent and specific heat"; chap. 6, pp. 173 – 215, "Some consequences of Black's work".

589 McKIE, D. On some MS. copies of Black's chemical lectures – VI.
Annals of Science 23 (1967) 1 – 33.
> On the MS notes of C. Blagden, N. Dimsdale, R. Dobson,

H. Richardson, BM MS Add. 52495, A. Anderson.

590 TALBOT, G.R. Origins and solutions of some problems in heat in the eighteenth century, chaps. 11 – 14.
(Ph.D. thesis, Victoria University of Manchester, 1967)
> Chap. 11, pp. 11.1 – 11.36, "Published work on specific and latent heats before 1784; evidence for Black's priority"; chap. 12, pp. 12.1 – 12.55, "Black, Watt and the discovery of specific and latent heats"; chap. 13, pp. 13.1 – 13.47, "The manuscript lectures"; chap. 14, pp. 14.1 – 14.18, "Robison's edition of Black's lectures and some early ideas about thermal conductivity".

591 GERSTNER, P.A. James Hutton's theory of the earth and his theory of matter.
Isis 59 (1968) 26 – 31.
> Suggests H.'s theory derives from interests in chemistry and latent heat, developed through his friendship with B.

592 ADVOCATUS DIABOLI MALLEATOR. A genius in hot water in Glasgow, with a Bicentenary reappraisal of some facts.
College Courant [Glasgow University] 21 (1969) 25 – 30.
> James Watt, with some references to B.

593 FRENCH, R.K. Robert Whytt, the soul, and medicine, chap. 2, pp. 17 – 26. London: The Wellcome Institute of the History of Medicine, 1969. (Publications of the . . . Institute . . . , new ser. vol. 17).
> "The controversy with Alston: Black and Whytt".

593a ROBINSON, E. James Watt and early experiments in alkali manufacture.
In: MUSSON, A.E. & ROBINSON, E., Science and technology in the Industrial Revolution, chap. 10, pp. 352 – 371. Manchester: University Press, 1969.

594 SWINBANK, P. James Watt and his shop.

Works on Black

Glasgow University Gazette 59 (1969) 5 – 8.
> Discussion of W.'s Waste Book, 1757 – 1763, and Ledger, 1764 – 1769, with references to work done for B. and for the Black, Wilson & Watt partnership.

594a SWINBANK, P. A University partnership. College Courant [Glasgow University] 21 (1969) 3.
> B.'s association with Watt at Glasgow.

595 VERBRUGGEN, F. Joseph Black en de antiphlogistische teorie.
Scientiarum Historia 11 (1969) 109 – 121.
> Very brief English summary on p. 121; further brief remarks on this paper, by C.E. Perrin, in *Ambix* 29 (1982) p. 171 note 15. Argues B. did not accept Lavoisier's antiphlogistic theory until 1790/1791.

596 DONOVAN, A.L. The origins of pneumatic chemistry: William Cullen, Joseph Black and the unification of natural philosophy. (Ph.D. thesis, Princeton University, 1970).
> In particular Part 3, chaps. 8 – 10, pp. 258 – 359, "The development of Joseph Black's theory of causticity". For Donovan's subsequent monograph see no. 609.

597 GUERLAC, H. Black, Joseph.
Dictionary of Scientific Biography, vol. 2 (1970) pp. 173 – 183.
> Reprinted, with one added comment, in no. 617.

598 ROBINSON, E. & McKIE, D. Partners in science: letters of James Watt and Joseph Black. Edited with introductions and notes by Eric Robinson and Douglas McKie. London: Constable, 1970. xvi, 502 p.
> American edition: Cambridge, Mass.: Harvard University Press, 1970.

599 SCHOFIELD, R.E. Mechanism and materialism: British

natural philosophy in an age of reason. Princeton: Princeton University Press, 1970.
<blockquote>pp. 185 – 190 and 219 – 226 for B.'s theories of heat and chemistry.</blockquote>

600 CARDWELL, D.S.L. From Watt to Clausius: the rise of thermodynamics in the early industrial age, pp. 34 – 57; 300 – 303. London: Heinemann, 1971.
<blockquote>Section entitled "The Scottish school": discusses B. and in particular the Black–Watt interaction at Glasgow University.</blockquote>

601 SMEATON, W.A. Some comments on James Watt's published account of his work on steam and steam engines.
Notes and Records of the Royal Society of London 26 (1971) 35 – 42.

602 DAVIS, A.B. & EKLUND, J.B. Magnesia alba before Black.
Pharmacy in History 14 (1972) 139 – 146.

603 EKLUND, J.B. & DAVIS, A.B. Joseph Black matriculates: medicine and magnesia alba.
Journal of the History of Medicine and Allied Sciences 27 (1972) 396 – 417.

604 MORRIS, R.J. Lavoisier and the caloric theory.
British Journal for the History of Science 6 (1972) 1 – 38.
<blockquote>In particular pp. 27 – 30, "Influence of Black".</blockquote>

605 ODDY, R. Some chemical apparatus blown by hand in the late 18th to early 19th century.
In: Annales du 5^e Congres International d'Etude Historique du Verre, Prague, 6 – 11 juillet 1970, pp. 225 – 231. Liege: Edition du Secretariat General [de l'Association Internationale pour l'Histoire du Verre],

1972.
> 25 items of glass chemical apparatus in the Royal Museum of Scotland, associated with B.

606 OLDROYD, D.R. Two little known copies of Black's lecture notes. Annals of Science 29 (1972) 35 – 37.
> The MS notes by A. Monro *tertius*, and R. Bellman.

607 LE GRAND, H.E. A note on fixed air: the universal acid. Ambix 20 (1973) 88 – 94.
> B. is mentioned only briefly, on p. 89.

608 LEICESTER, H.M. Development of biochemical concepts from ancient to modern times, pp. 128 – 129. Cambridge, Mass.: Harvard University Press, 1974. (Harvard monographs in the history of science).
> Brief discussion of the first part of B.'s *Dissertatio medica inauguralis*.

609 DONOVAN, A.L. Philosophical chemistry in the Scottish Enlightenment: the doctrines and discoveries of William Cullen and Joseph Black. Edinburgh: University Press, 1975. x, 343 p.
> Includes many excerpts from Black MSS.
> Reviewed by P.M. Heimann in *British Journal for the History of Science* 9 (1976) 328 – 329; by W.P.D. Wightman in *Ambix* 24 (1977) 65; by D.M. Knight in *Annals of Science* 33 (1976) 509 – 510; in *Medical History* 21 (1977) 324 – 325; by R.E. Schofield in *Technology and Culture* 18 (1977) 528 – 529; by J. Starobinski in *Gesnerus* 35 (1978) 163 – 164; etc.
> Based on Donovan's thesis, for which see no. 596.

609a LINDSAY, R.B. Energy: historical development of the concept, pp. 175 – 176, 190 – 203. Stroudsburg, Pennsylvania: Dowden, Hutchinson & Ross, Inc., 1975. (Benchmark papers on energy /1).

Works on Black

Very brief historical note followed by reprint of a passage from *Lectures on the elements of chemistry*, Philadelphia, 1807 (The nature of heat).

610 ANDERSON, R.G.W. & SIMPSON, A.D.C. Edinburgh & medicine. A commemorative catalogue of the exhibition held at the Royal Scottish Museum, Edinburgh, June 1976 — January 1977 . . . Edinburgh: The Royal Scottish Museum, 1976.

In particular the following exhibits: No. 184, De humore acido, 1754 (p. 40); No. 185, Black's balance (p. 40; fig. 10); No. 186, Alembic (p. 40); No. 187, Cucurbit (p. 40); No. 188, Bottle with ground glass stopper (p. 40); No. 189, Joseph Black [etching by John Kay] (p. 40; fig. 9).

611 COWEN, D.L. The Edinburgh Pharmacopoeia. In: ANDERSON, R.G.W. & SIMPSON, A.D.C., eds., The early years of the Edinburgh Medical School, pp. 25 — 45. Edinburgh: The Royal Scottish Museum, 1976.

B.'s service on the Revision Committee and his hand in the 1774, 1783, and 1792 editions are briefly described on pp. 28, 34.

612 DONOVAN, A.L. Pneumatic chemistry and Newtonian natural philosophy in the eighteenth century: William Cullen and Joseph Black.
Isis 67 (1976) 217 — 228.

613 EKLUND, J. Of a spirit in the water: some early ideas on the aerial dimension.
Isis 67 (1976) 527 — 550.

614 GUERLAC, H. Chemistry as a branch of physics: Laplace's collaboration with Lavoisier.
Historical Studies in the Physical Sciences 7 (1976) 193 — 276.

Footnote 140, pp. 250 — 251, disproves the attribution of

the invention of the ice calorimeter to B.

615 MACKIE, A. A famous foursome and its influence on science and technology. Chemistry and Industry (7 Feb. 1976) 98 – 103.
Joseph Black, Archibald Cochrane, James Hutton, James Watt. In particular pp. 102 – 103 for B.

615a PANCALDI, G. Joseph Black: la chimica tra newtonianesimo e rivoluzione industriale. In: SANTUCCI, A., ed., Scienza e filosofia scozzese nell'eta di Hume, pp. 209. Bologna: Il Mulino, 1976.

616 PRATT, H.T. Samuel L. Mitchill's evaluation of the lectures of Joseph Black. Journal of Chemical Education 53 (1976) 745 – 746.

617 GUERLAC, H. Joseph Black. In: GUERLAC, H., Essays and papers in the history of modern science, pp. 285 – 303. Baltimore; London: The Johns Hopkins University Press, 1977.
Reprinted, with one added comment on p. 300, from no. 597.

618 SEBASTIANI, F. Black e le teorie sulla natura del calore. Giornale di Fisica 18 (1977) 304 – 312.

618a WITTEN, L.C. & PACHELLA, R. Alchemy and the occult. A catalogue of books and manuscripts from the collection of Paul and Mary Mellon given to Yale University Library, vol. 4, Manuscripts, 1675 – 1922. New Haven: Yale University Library, 1977.
Describes and illustrates the three Mellon Collection sets of Black lecture notes: MS 113, pp. 658 – 661 (illus. on pp. 659 – 660); MS 118, pp. 678 – 681 (illus. on pp. 679 – 680); MS 125, pp. 705 – 707 (illus. on p. 706).

619 ANDERSON, R.G.W. The Playfair Collection and the

teaching of chemistry at the University of Edinburgh 1713 – 1858. Edinburgh: The Royal Scottish Museum, 1978. viii, 175 p.
> In particular chap. 2, pp. 19 – 33, "Joseph Black"; and pp. 63 – 159, "Catalogue of the Playfair Collection", describing many items associated with B.

620 DONOVAN, A.L. James Hutton, Joseph Black and the chemical theory of heat.
Ambix 25 (1978) 176 – 190.

620a RON, M. From alchemy to atoms. An exhibition of books, documents, Mss., etc. on the history of chemistry and chemical technology from the Sidney M. Edelstein Collection, p. 31. Jerusalem: Jewish National and University Library, 1978.
> Describes the Edelstein Collection set of Black lecture notes.

621 DONOVAN, A.L. Scottish responses to the new chemistry of Lavoisier.
Studies in Eighteenth Century Culture 9 (1979) 237 – 249.

622 DONOVAN, A.L. Towards a social history of technological ideas: Joseph Black, James Watt, and the separate condenser.
In: BUGLIARELLO, G. & DONER, D.B., eds., The history and philosophy of technology, pp. 19 – 30. Urbana; London: University of Illinois Press, 1979.

623 SHACKLETON, R. John Black and Montesquieu – the search for a correspondence.
In: WELLECK, R. & RIBEIRO, A., eds., Evidence in literary scholarship: essays in memory of James Marshall Osborn, pp. 215 – 227. Oxford: Clarendon Press, 1979.
> Prints on pp. 222 – 225 three letters from Montesquieu to John Black, 1749 – 1751, with Joseph Black's covering note.

624 BUCHANAN, W.W. & BROWN, D.H. Joseph Black (1728 – 1799): Scottish physician and chemist. Practitioner 224 (1980) 663 – 666.

625 TREASE, G.E. Joseph Black (1728 – 1799). British Journal of Pharmaceutical Practice, vol. 1 no. 10 (Feb. 1980) 32 – 33.

626 DUNCAN, A.M. Styles of language and modes of chemical thought.
Ambix 28 (1981) 83 – 107.
> In particular pp. 85, 89 – 90 for B.'s views on "affinity"/"attraction", and his definition of "chemistry".

627 EMERSON, R.L. The Philosophical Society of Edinburgh, 1748 – 1768. British Journal for the History of Science 14 (1981) 133 – 176.

628 MELDRUM, A.N. The eighteenth century revolution in science – the first phase.
In: MELDRUM, A.N. Essays in the history of chemistry. New York: Arno Press, 1981. (The development of science).
> Reprinted from no. 501.
> N.B. this reprint collection does not have its own pagination, but simply reproduces the various paginations of the original papers.

629 PARTINGTON, J.R. & McKIE, D. Historical studies on the phlogiston theory. – I. The levity of phlogiston.
In: PARTINGTON, J.R. & McKIE, D., Historical studies on the phlogiston theory. New York: Arno Press, 1981. (The development of science)
> Reprinted from no. 517.
> N.B. this reprint collection does not have its own pagination, but simply reproduces the various paginations of the original papers.

Works on Black

630 SEBASTIANI, F. La memoria voltiana intorno al calore. Physis 23 (1981) 89 – 113.
In particular section II, pp. 90 – 95, for B. Very brief English summary on p. 113, not mentioning B.

631 ANDERSON, R.G.W. Joseph Black: an outline biography.
In: SIMPSON, A.D.C., ed., Joseph Black. A commemorative symposium, pp. 6 – 11. Edinburgh: The Royal Scottish Museum, 1982.

632 BUCHANAN, P.D. Quantitative measurement and the design of the chemical balance 1750 – c1900, pp. 194 – 196. (Ph.D. thesis, Imperial College, 1982).
Brief discussion of the sensitivity and maximum load of B.'s balance, and the availability of suitable wire for making the weights by Lewis's method; B.'s balance in the Royal Museum of Scotland is not among those analysed in detail by the author.

633 CHRISTIE, J.R.R. Joseph Black and John Robison.
In: SIMPSON, A.D.C., ed., Joseph Black. A commemorative symposium, pp. 47 – 52. Edinburgh: The Royal Scottish Museum, 1982.

634 COLE, W.A. Manuscripts of Joseph Black's lectures on chemistry.
In: SIMPSON, A.D.C., ed., Joseph Black. A commemorative symposium, pp. 53 – 69. Edinburgh: The Royal Scottish Museum, 1982.
Census of the 97 currently known manuscripts of the lectures (including 8 unlocated manuscripts). Each entry gives: Repository, call-mark, and date; Title; Physical description; Lectures included; Names of compiler, owners, etc.; Published references. Manuscripts subsequently coming to light include the set sold at auction by Christie, Manson & Woods in their sale "Books from the library of Matthew Boulton and his family", 12th Dec. 1986 (Lot 24; sale

catalogue p. 16, including illus. of one page).

635 DOIG, A. Dr Black, a remarkable physician.
In: SIMPSON, A.D.C., ed., Joseph Black. A commemorative symposium, pp. 37 − 41. Edinburgh: The Royal Scottish Museum, 1982.

636 DONOVAN, A.L. William Cullen and the research tradition of eighteenth-century Scottish chemistry.
In: CAMPBELL, R.H. & SKINNER, A.S., eds., The origins and nature of the Scottish Enlightenment, pp. 98 − 114. Edinburgh: John Donald Publishers Ltd, 1982.

637 DOYLE, W.P. Black, Hope and Lavoisier.
In: SIMPSON, A.D.C., ed., Joseph Black. A commemorative symposium, pp. 43 − 46. Edinburgh: The Royal Scottish Museum, 1982.

638 GUERLAC, H. Joseph Black's work on heat.
In: SIMPSON, A.D.C., ed., Joseph Black. A commemorative symposium, pp. 13 − 22. Edinburgh: The Royal Scottish Museum, 1982.

639 LAWRENCE, C. Joseph Black: the natural philosophical background.
In: SIMPSON, A.D.C., ed., Joseph Black. A commemorative symposium, pp. 1 − 5. Edinburgh: The Royal Scottish Museum, 1982.

640 PERRIN, C.E. A reluctant catalyst: Joseph Black and the Edinburgh reception of Lavoisier's chemistry. Ambix 29 (1982) 141 − 176.

641 SIMPSON, A.D.C. Joseph Black 1728 − 1799. A commemorative symposium. Papers presented at a symposium held in the Royal Scottish Museum on 4 November 1978 in association with The Scottish Society of the History of Medicine, together with a survey of

Works on Black

manuscript notes of Joseph Black's lectures on chemistry. Edited by A.D.C. Simpson. Edinburgh: The Royal Scottish Museum, 1982. viii, 69 p. (Royal Scottish Museum studies).
> The individual contributions are separately listed in the present bibliography.

642 SWINBANK, P. Experimental science in the University of Glasgow at the time of Joseph Black.
In: SIMPSON, A.D.C., ed., Joseph Black. A commemorative symposium, pp. 23 – 35. Edinburgh: The Royal Scottish Museum, 1982.

643 DOYLE, W.P. Joseph Black.
University of Edinburgh Journal 31 (1983) 40 – 42.

644 PERRIN, C.E. Joseph Black and the absolute levity of phlogiston.
Annals of Science 40 (1983) 109 – 137.

645 JOHNSTONE, A.H. Joseph Black – the father of chemical education?
Journal of Chemical Education 61 (1984) 605 – 606.

646 JONES, J. The geological collection of James Hutton.
Annals of Science 41 (1984) 223 – 244.
> Many references to B.; prints on pp. 224 – 225 his letter to the Royal Society of Edinburgh offering them H.'s collection.

647 STANSFIELD, D.A. Thomas Beddoes M.D., 1760 – 1808. Dordrecht: D. Reidel Publishing Company, 1984. xx, 306 p.
> Many references to Black; in particular: an extract from an undated letter from Beddoes to C.B. Trye briefly evaluating the lectures of A. Monro *secundus*, Black, and Cullen (pp. 23 – 24); and quotations from the Beddoes–Black correspondence (in chap. 4, pp. 31 – 59).

Works on Black

648 ANDERSON, R.G.W. Black, Joseph.
In: PORTER, R., ed., Dizionario Biografico della Storia della Medicina e delle Scienze Naturali (Liber Amicorum), tomo 1, pp. 112 – 113. Milano: Franco Maria Ricci, 1985.

649 APPLEBY, J.H. John Grieve's correspondence with Joseph Black and some contemporaneous Russo-Scottish medical intercommunication.
Medical History 29 (1985) 401 – 413.

650 MACKENZIE, R.C. & PROKS, I. Comenius and Black: progenitors of thermal analysis.
Thermochimica Acta 92 (1985) 3 – 14.

651 SEBASTIANI, F. La fisica dei fenomeni termici nella seconda meta del Settecento: le teorie sulla natura del calore da Black a Volta.
Physis 27 (1985) 45 – 126.

652 ANDERSON, R.G.W. Joseph Black.
In: DAICHES, D., JONES, P. & JONES, J., eds., A hotbed of genius: the Scottish Enlightenment 1730 – 1790, pp. 92 – 114. Edinburgh: University Press, 1986.

653 UNIVERSITY OF EDINBURGH, INSTITUTE FOR ADVANCED STUDIES IN THE HUMANITIES. 'A hotbed of genius': the Scottish Enlightenment 1730 – 90. Check-list of exhibits presented with the compliments of the sponsors The Royal Bank of Scotland.
[Edinburgh, 1986.]
> Catalogue of the exhibition held at the Royal Museum of Scotland, 8 July – 20 September 1986. In particular section B. Black, pp. 13 – 15, exhibits 3.b.1 – 3.b.33.

Part V

ICONOGRAPHY

901 COCHRANE, T. [Joseph Black] [Pencil drawing] [1767 or 1768] ca. 3.8 x 2.0 cm.
    Sketch in vol. 1 p. 622 of C.'s MS lecture notes (Strathclyde University, Andersonian Library).
    Head and shoulders, facing left, profile.
    Reproduced in McKIE, D., ed., *Thomas Cochrane. Notes from Doctor Black's lectures on chemistry 1767/8*, unnumbered pl. at end. Wilmslow, 1966.

902 COCHRANE, T. [Joseph Black lecturing, with chemical flask] [Ink drawing] [1767 or 1768] ca. 11.9 x 8.5 cm.
    Sketch in vol. 2 p. 3 of C.'s MS lecture notes (Strathclyde University, Andersonian Library).
    Half length, facing left, head in profile, standing lecturing at a table, a chemical flask raised in his right hand, his left hand resting on his lecture notes; behind him is another figure, half length, full face.
    Reproduced in McKIE, D., ed., *Thomas Cochrane. Notes from Doctor Black's lectures on chemistry 1767/8*, p. 108. Wilmslow, 1966.

903 COCHRANE, T. [Joseph Black lecturing, in anatomical theatre] [Ink drawing] [1767 or 1768] ca. 15.9 x 9.5 cm.
    Sketch in vol. 2 p. 682 of C.'s MS lecture notes (Strathclyde University, Andersonian Library).
    Half length, facing left, head in profile, standing lecturing at a table, both hands resting on his lecture notes; to the right are three items of chemical apparatus on the table; to the upper left the audience is roughly indicated. Adjacent to B.'s head is a less finished sketch of another head.
    Reproduced in McKIE, D., ed., *Thomas Cochrane. Notes from Doctor Black's lectures on chemistry 1767/8*, p. xxviii. Wilmslow, 1966. (The experiment for which the apparatus was used is described on pp. xviii – xix.)

Iconography

904 MARTIN, D. [Joseph Black] [Oil painting] [ca. 1770] 76 x 62 cm.
Edinburgh University (RICE, D.T. & McINTYRE, P., *The University portraits*, pp. 15 – 16; pl. 11. Edinburgh: Edinburgh University Press, 1957.)
Half length, facing left, right hand resting on table. Reproduced in *Isis* 48 (1957) pl. facing p. 132 (suggesting a dating of ca. 1766); and in McKIE, D., ed., *Thomas Cochrane. Notes from Doctor Black's lectures on chemistry 1767/8*, p. ii. Wilmslow, 1966.

905 ANON. Professor Joseph Black, M.D. chemist. [Chalk drawing] [ca. 1780] 44 x 57 cm.
Scottish National Portrait Gallery (THOMSON, D. & LOCKHART, S.B., *Scottish National Portrait Gallery concise catalogue*, p. 12. Edinburgh: The Trustees of the National Galleries of Scotland, 1977.)
Head and shoulders, facing left; in an oval. Reproduced in *Journal of Chemical Education* 11 (1934) 485; and in SIMPSON, A.D.C., ed., *Joseph Black 1728 – 1799. A commemorative symposium*, p. vi. Edinburgh, 1982.

906 KAY, J. [Joseph Black walking] [Etching] JK. fect. 1787. 7.9 x 5.4 cm.
Full length, facing left, walking, in profile, a stick in his right hand; on the left a rock face whose outline forms the profiles of seven of his contemporaries at Edinburgh University (cf Kay's similar portrait of J. Hutton). Reissued in the collection *A series of original portraits and caricature etchings, by the late John Kay*, vol. 1, facing p. 52. Edinburgh, 1838 (later issues 1842, 1877). Reproduced, enlarged, in SIMPSON, A.D.C., ed., *Joseph Black 1728 – 1799. A commemorative symposium*, p. 6. Edinburgh, 1982.

907 KAY, J. [Joseph Black lecturing] [Etching] JK. fect. 1787. 10.5 x 8.6 cm.
Half length, facing left, standing lecturing at a table, a slip

Iconography

of paper in his right hand and his spectacles in his left; on the table are more slips and two books; apparatus for preparing CO_2, with a bird cage and a candle for demonstrating its properties; and a geological hammer (cf Martin's 1787 painting (no. 909)). Reissued in the collection *A series of original portraits and caricature etchings, by the late John Kay*, vol. 1, facing p. 54. Edinburgh, 1838 (later issues 1842, 1877).
Reproduced, enlarged, in ANDERSON, R.G.W., *The Playfair Collection and the teaching of chemistry at Edinburgh University 1713 – 1858*, p. 25. Edinburgh, 1978.

908 KAY, J. Philosophers. [Etching] JK. fect. 1787. 7.9 x 11.4 cm.

Half length, facing left, seated, his right arm resting on a pile of books and his right hand thrust inside his coat; facing James Hutton. Reissued in the collection *A series of original portraits and caricature etchings, by the late John Kay*, vol. 1, facing p. 56. Edinburgh, 1838 (later issues 1842, 1877).
Reproduced in READ, J., *Humour and humanism in chemistry*, fig. 45, facing p. 167. London, 1947; and in CROWTHER, J.G., *Scientists of the Industrial Revolution*, pl. 2, facing p. 57. London, 1962.

908a LORENZ, M. Joseph Black 1728 – 1978. [Medallion] ML [after Kay, [Joseph Black lecturing]] [19]78. 16 cm.

Half length, facing left, a slip of paper in his right hand and his spectacles in his left; in the background is a diagram of elective attractions.

909 MARTIN, D. [Joseph Black] [Oil painting] 1787. 127 x 102 cm.

Royal Medical Society, Edinburgh, on loan to the Scottish National Portrait Gallery (THOMSON, D. & LOCKHART, S.B., *Scottish National Portrait Gallery concise catalogue*, p. 12. Edinburgh: The Trustees of the National Galleries

Iconography

of Scotland, 1977.) Three quarter length, facing left, standing lecturing at a table, a U tube raised in his right hand; on the table are his spectacles and lecture notes, apparatus for preparing CO_2, and a bird cage and candle for demonstrating its properties (cf Kay's etching (no. 907)).
Reproduced in colour in DAICHES, D., JONES, P. & JONES, J., eds., *A hotbed of genius,* p. 102. Edinburgh, 1986.

909a BARR, J. [Joseph Black] [Oil painting] [after Martin (1787)] [186–?] ca. 75 x 60 cm.
Royal College of Physicians and Surgeons of Glasgow (GIBSON, T., *The Royal College of Physicians and Surgeons of Glasgow: a short history based on the portraits and other memorabilia,* pp. 80 – 83 (with colour reproduction on p. 80). Loanhead: Macdonald Publishers (Edinburgh), 1983.); the MS minutes of the College for Aug. 1868 briefly describe the origin and acquisition of this painting.
Head and shoulders, facing left.

910 RAEBURN, H. [Joseph Black] [Oil painting] [ca. 1788] 124 x 101 cm.
Glasgow University; purchased at Christie's sale "Important English pictures", 18th April 1986 (lot 129A; sale catalogue pp. 172 – 173, including colour reproduction, list of exhibitions, and literature).
Three quarter length, facing left, seated; dark dress with white lace at throat and wrists; red chair and curtain.

911 HEATH, J. Joseph Black, M.D. F.R.S.E. [Stipple engraving] Engraved by Jas. Heath from a picture by Raeburn. [London] : Published . . . by J. Heath . . . & J.P. Thompson, 1800. 20.2 x 16.9 cm.
In another issue the title continues: Late Professor of Chemistry in the University of Edinburgh.
Half length, facing left, seated; in an oval.
Reproduced in LAW, R.J., *James Watt and the separate*

Iconography

condenser, p. 6. London: HMSO, 1969.

912 ANON. Dr. Joseph Black. [Stipple engraving] [after Raeburn] [London] : Published by Longman & Rees, 1803. 11.1 x 8.9 cm.

> Frontispiece to *Lectures on the elements of chemistry . . . by the late Joseph Black . . . now published . . . by John Robison . . .* , vol. 1. Edinburgh, 1803. "The portrait of Dr. Black prefixed to these lectures is an excellent likeness." – T. Thomson in *The history of chemistry*, vol. 1, p. 334, footnote. London, 1830.
> Half length, facing left, seated; in an oval.

913 ROSMASLER, J. Dr. Joseph Black. [Stipple engraving] Rosmasler. sculp: [after Raeburn] [Hamburg: Hoffmann, 1804] 9.3 x 6.7 cm.

> Plate to *D. Josef Black's . . . Vorlesungen uber die Grundlehren der Chemie . . . Aus dem Englischen ubersetzt . . . von D. Lorenz von Crell*, Bd. 1. Hamburg, 1804.
> Half length, facing right, seated; in an oval.

914 KNEASS, W. Dr. Joseph Black. [Stipple engraving] Grav'd by W. Kneass [after Raeburn] Philad.: pub. by Matthew [sic] Carey, 1807. 10.6 x 8.2 cm.

> Frontispiece to *Lectures on the elements of chemistry . . . by the late Joseph Black . . .* , first American . . . edition, vol. 1. Philadelphia, 1807.
> Half length, facing left, seated; in an oval.

915 DEAN, T.A. Joseph Black, M.D. F.R.S.E. [Line and stipple engraving] Raeburn. pinxt., Dean. sculpt. London: Published by Henry Colburn & Richard Bentley, 1830. 9.8 x 7.1 cm.

> Frontispiece to THOMSON, T., *The history of chemistry*, vol. 1. London, 1830.
> Half length, facing left, seated.

916 POSSELWHITE, J. Black. [Stipple engraving] Engraved

Iconography

by J. Posselwhite. From a print by Jas. Heath, after a picture by Raeburn. London: Published by Charles Knight & Co., [183–] 11.8 x 8.6 cm.

Another issue is described as: Under the superintendence of the Society for the Diffusion of Useful Knowledge.

Plate to KNIGHT, C., *The gallery of portraits, with memoirs*; and to CRAIK, G.L., *The pictorial history of England during the reign of George the Third*; and to BROUGHAM, H.P., *Lives of men of letters and science*, facing p. 324. London, 1845.

Half length, facing left, seated.

917 ROGERS, J. Joseph Black, M.D. Professor in the University of Edinburgh. [Stipple engraving] J. Rogers [after] Sir Henry Raeburn. Glasgow: Blackie & Son, [18——] 12.3 x 9.6 cm.

Plate to CHAMBERS, R., *A biographical dictionary of eminent Scotsmen*.

Three quarter length, facing left, seated.

918 COOK, C. Dr. Black. Professor of Chemistry, Glasgow & Edinburgh. [Stipple engraving] C. Cook. [after] Raeburn. Glasgow [etc.] : William Mackenzie, [185–?] 14.0 x 10.8 cm.

Plate to MUSPRATT, S., *Chemistry theoretical, practical & analytical*, division 4. [Glasgow [etc.]: William Mackenzie, 185–?] (Muspratt's *Chemistry* was published in seven divisions, constituting two volumes, [1853 – 1861]; division 4 is the first division of vol. 2.)

Half length, facing left, seated.

An Autotype copy of this engraving appears as a plate to the *Asclepiad*, vol. 7 no. 27 (July 1890).

919 SCHARF, G. [Joseph Black] [Lithograph] [after Raeburn]

Bust, facing right.

920 COOPER, J. Dr. Joseph Black. [Wood engraving]

Iconography

J. Cooper sc. [after Raeburn] [London: John Murray, 1865] 10.2 x 8.4 cm.
> Illustration to SMILES, S., *Lives of Boulton and Watt. Principally from the original Soho MSS*, p. 132. London: John Murray, 1865.
> Half length, facing left, seated; in an oval.

921 ANON. Joseph Black, M.D., 1728 — 1799. [Oil painting] [after Raeburn] [19—?] 76 x 61 cm.
> New York Academy of Medicine (*Portrait catalog of the New York Academy of Medicine*, first supplement, 1959 — 1965, p. 68. Boston: G.K. Hall & Co., 1965)

922 TASSIE, J. Joseph Black M.D. [Wax medallion]
> The Wedgwood Museum, Barlaston, has a wax; the Victoria and Albert Museum others; another, in the possession of E.C. Watson, is described and illustrated in the *American Journal of Physics* 22 (1954) 32.

923 TASSIE, J. Joseph Black M.D. [Glass-paste medallion] Tassie F. 1788. 7.3 cm.
> (GRAY, J.M., *James and William Tassie*, no. 37, p. 87. Edinburgh: Walter Greenoak Patterson, 1894.)

924 TASSIE, J.
> "A smaller version of no. 37 (about 12/16 in.) is stated to have formed part of Lot No. 45 in the Vernon Sale." — GRAY, J.M., *James and William Tassie*, no. 38, p. 87. Edinburgh: Walter Greenoak Patterson, 1894.

925 TASSIE, J. [Joseph Black] [Wedgwood medallion] Tassie F. 1788. 6.8 cm.
> (REILLY, R. & SAVAGE, G., *Wedgwood: the portrait medallions*, p. 64 [including illus.]. London: Barrie & Jenkins Ltd, 1973.)

926 MACKENZIE, K. [Joseph Black] [Stipple engraving] 7.8 x 6.5 cm.

Iconography

Plate to the *Philosophical Magazine* — one of the "heads of philosophers deceased" produced for the "embellishment" of the magazine (see the note to the binder in vol. 9 (1801) p. 382).
Bust, facing left, profile; in an oval.

927 WATT, J. In the period 1806 — 1814 Watt produced many copies of portrait medallions etc. using his "sculpturing machines". The machines, the work done on them, and associated material are now in the Science Museum, London (Inventory no. 1926 - 1075). For a study of the machines see DICKINSON, H.W., *The garret workshop of James Watt*, pp. 14 — 24; figs. 4 — 6. London: HMSO, 1929.
The collection includes 25 versions of the Tassie/ Wedgwood medallion of Black (subcollection Moulds, casts, and material, nos. 46 — 70). At least six of these have been produced directly on the sculpturing machines, and are described below; their interest is more mechanical than artistic. The remainder are apparently moulds or casts in plaster or lead.

No. 49. Dr Black [Copy in alabaster] 1807. 3.2 cm within a slightly raised oval 4 cm, on a disc 5 cm dia. [Title and date from Watt's MS label on the back, which also bears the abbreviation Scr. [i.e. Scraping tool]].

No. 50. Dr Black [Copy in wood] 1808. 3.3 cm on a rectangle 6.2 x 5.8 cm. [Title and date from Watt's MS label on the back, which also bears the abbreviation Scr. [i.e. Scraping tool]].

No. 67. Black. [Copy in wood] Decr 1809. 6.8 cm on an oval 8.5 cm. [Title and date from Watt's MS written directly on the back].

Nos. 68 — 69. [Two other copies in wood closely resembling no. 67, but without titles, dates or other annotation].

No. 70. [Copy in alabaster] 7.4 cm on an oval 9.6 cm. [Without title, date or other annotation].

928 STANIER, R. Wm. Black, M.D. [Stipple engraving] Stanier sculpt. [London] : Published by J. Sewell, 1790. 12.7 x 9.5 cm.

> Plate to the *European Magazine* 22 (1792) 83 – 84, facing p. 84. "By a mistake which we are sorry for, though we are unable to assign any reason for the accident, the name of this gentleman in the copper-plate is erroneously called William, instead of Joseph, as it ought to be." – footnote on p. 83.
> Head and shoulders, facing left; in an oval within a rectangle.
> Reproduced in SMITH, E.F., *Old chemistries*, pl. facing p. 78. New York; London: McGraw-Hill Book Company, Inc., 1927.

929 BEUGO, J. [Joseph Black] [Stipple engraving] [not later than 1793] 7.9 x 6.4 cm.

> (BROMLEY, H., *A catalogue of engraved British portraits*, p. 383. London: Printed for T. Payne [etc.], 1793.)
> Bust, facing left, profile; in an oval.

930 ROBERTSON, – Joseph Black M.D. Professor of Chemistry in the University of Edinburgh. [Line engraving] Robertson sculp. [after] Thornton [in Greek script] [not later than 1793] 15.2 x 13.0 cm.

> (BROMLEY, H., *A catalogue of engraved British portraits*, p. 383. London: Printed for T. Payne [etc.], 1793.)
> Half length, facing right; in an oval within a rectangle.
> Reproduced in McKIE, D., ed., *Thomas Cochrane. Notes from Doctor Black's lectures on chemistry 1767/8*, p. xi. Wilmslow, 1966.

931 RIEDEL, C.F. D. Joseph Black, einst Professor der Chemie zu Edinburgh. [Stipple engraving] C.F. Riedel

Iconography

sc. Lips., [1801?] 8.7 x 7.1 cm.
Plate to *Allgemeines Journal der Chemie* 6 (1801).
Half length, facing right, profile; in an oval within a rectangle.
Reproduced in *Journal of Chemical Education* 13 (1936) 503.

932 READ, W. Dr. Black. [Stipple engraving] W. Read sc. 5.8 x 6.4 cm.
Half length, facing left.

JOSEPH BLACK

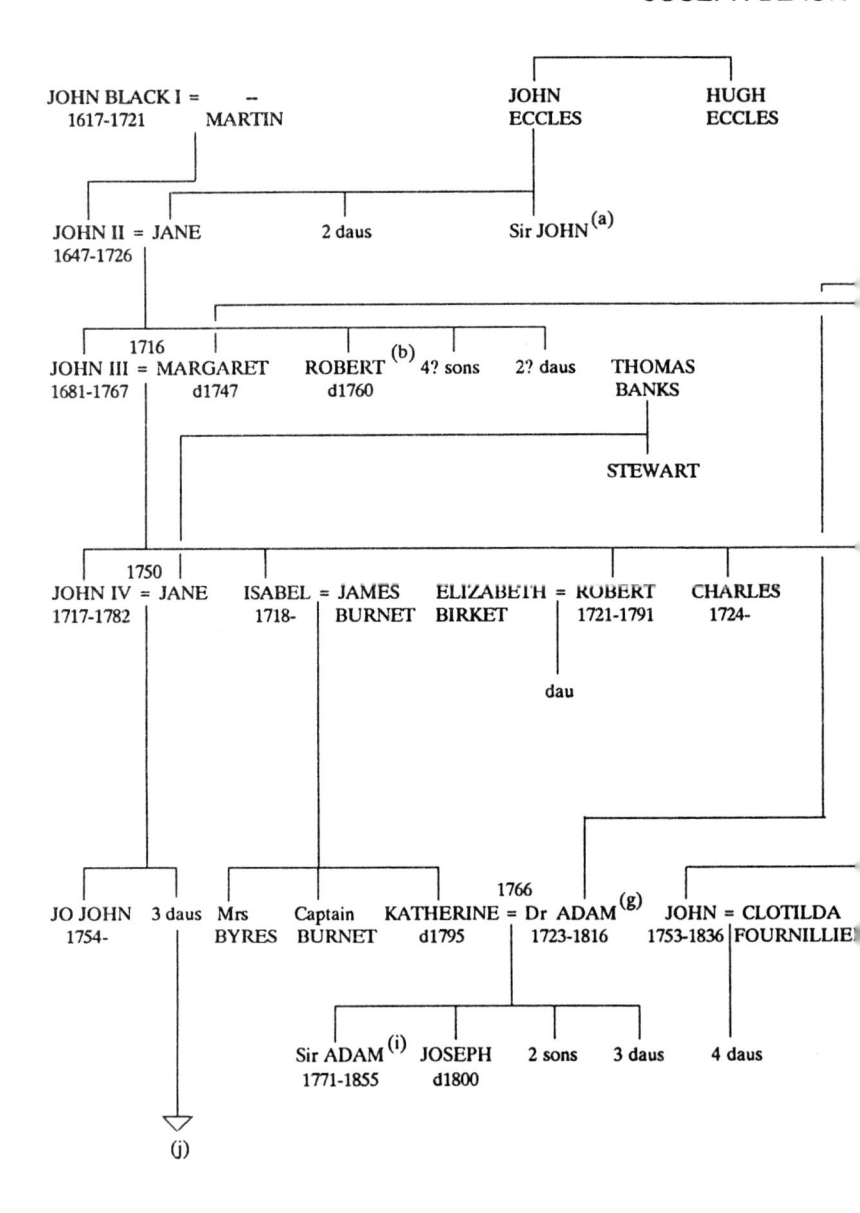

```
JOHN BLACK I  =   --              JOHN              HUGH
  1617-1721      MARTIN           ECCLES            ECCLES
      │                             │
      ├──────────────┐              ├──────────────┐
      │              │              │
  JOHN II = JANE   2 daus       Sir JOHN (a)
  1647-1726
      │
      ├──────────────────────────────────────────────────┐
      │                                                  │
      │    1716
  JOHN III = MARGARET    ROBERT (b)  4? sons   2? daus   THOMAS
  1681-1767    d1747      d1760                          BANKS
      │                                                    │
      │                                                  STEWART
      │
      ├──────────┬──────────────┬────────────────────┬──────────┐
      │ 1750     │              │                    │          │
  JOHN IV = JANE   ISABEL = JAMES   ELIZABETH = ROBERT    CHARLES
  1717-1782        1718-   BURNET   BIRKET     1721-1791  1724-
                                       │
                                      dau

      ├────────┬────────┬──────────┬─────────1766──────────┬─────────────────┐
      │        │        │          │                       │                 │
  JO JOHN  3 daus   Mrs       Captain   KATHERINE = Dr ADAM (g)  JOHN = CLOTILDA
  1754-            BYRES      BURNET    d1795       1723-1816    1753-1836 FOURNILLIE
                                            │
                              ┌─────────┬───┴────┬─────────┐
                          Sir ADAM (i)  JOSEPH   2 sons   3 daus    4 daus
                          1771-1855     d1800
    ▽
    (j)
```

See page 118 for Notes.

FAMILY TREE

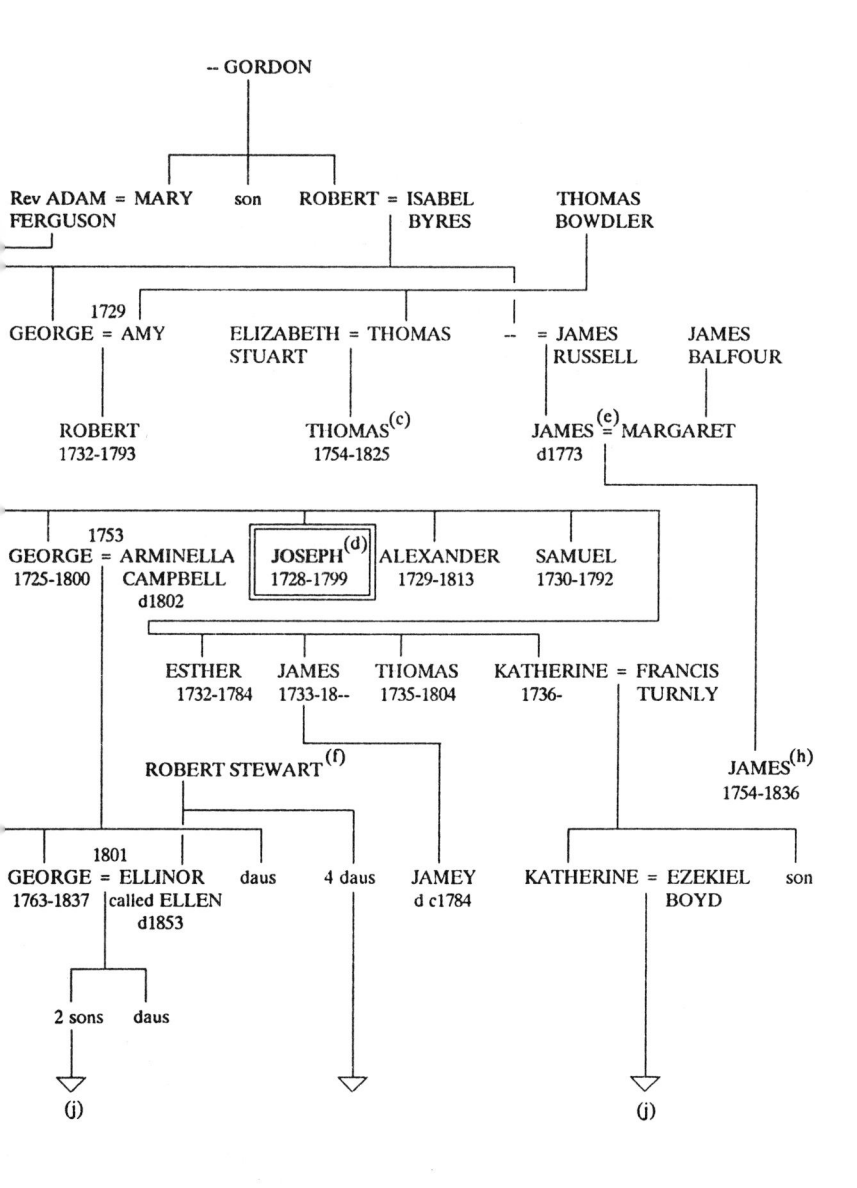

```
                                    -- GORDON
                                        |
        ┌───────────────────────────────┴──────┐
Rev ADAM = MARY    son    ROBERT = ISABEL            THOMAS
FERGUSON                          BYRES              BOWDLER
    |                       |
    |   1729                |                                   |
    ├─GEORGE = AMY     ELIZABETH = THOMAS      -- = JAMES    JAMES
    |                              STUART            RUSSELL  BALFOUR
    |                                   |                 |
    |   ROBERT                     THOMAS⁽ᶜ⁾         JAMES ⁽ᵉ⁾= MARGARET
    |   1732-1793                  1754-1825         d1773        |
    |                                                             |
    |  1753                                                       |
    ├─GEORGE = ARMINELLA   │JOSEPH⁽ᵈ⁾│ ALEXANDER   SAMUEL         |
    | 1725-1800  CAMPBELL  │1728-1799│ 1729-1813   1730-1792      |
    |            d1802                                            |
    |                                                             |
    |        ESTHER    JAMES     THOMAS    KATHERINE = FRANCIS
    |        1732-1784 1733-18-- 1735-1804 1736-       TURNLY
    |                                                      |
    |           ROBERT STEWART⁽ᶠ⁾                        JAMES⁽ʰ⁾
    |                                                    1754-1836
    |   1801
    ├─GEORGE = ELLINOR   daus   4 daus   JAMEY    KATHERINE = EZEKIEL  son
    | 1763-1837 called ELLEN            d c1784               BOYD
    |           d1853
    |
 ┌──┴──┐
 2 sons  daus
   ▽               ▽                                  ▽
  (j)             (j)                                (j)
```

117

NOTES ON FAMILY TREE

a) Lord Mayor of Dublin

b) HM Consul in Cadiz

c) Editor of *The Family Shakespeare*

d) Joseph was the fifth son and ninth child in a family of nine sons and six daughters. John IV was the eldest, and Katherine the youngest, child. Three daughters born 1719-1723 are not included in this family tree.

e) Professor of Natural Philosophy, Edinburgh University

f) For the Stewarts of Ballydrain see Riddell (item 489).

g) Professor of Natural Philosophy and subsequently Professor of Moral Philosophy, Edinburgh University

h) First Professor of Clinical Surgery, Edinburgh University

i) Keeper of the Regalia in Scotland

j) For some descendants see Riddell (item 489).

NAME INDEX

The references in the index are to entries, not pages

Académie Royale des Sciences (Paris) 404
Academy of Sciences (St Petersburg) 403
Accum, F 266
Addison, W I 485
Advocatus Diaboli Malleator 592
Allan, E 250 cf Annan, E
Alston, C 201, 202, 204, 593
Anderson, A 589
Anderson, C 429
Anderson, R G W 610, 619, 631, 648, 652, 907
Andersonian Library 901, 902, 903
Andrews, T 459, 460, 492
Annan, E 262 cf Allan, E
Appleby, J H 649
Arago, D F 441

Baldinger, E G 303
Barclay, A 494
Barnett, M K 528, 529
Barr, J 909a
Beaufoy, H B F 212
Beddoes, T 240, 251, 647
Bell, J 504
Bell Museum of Pathobiology 6
Bellman, R 606
Bergman, T 518
Beugo, J 929
Biggs, B 251
Bishop, T H 553
Black, A 573
Black, G 478
Black, James 454
Black, John 623
Blagden, C 231, 589
Blair, J 402
Bloxam, T 451
Boehm, M F 304
Boswell, J 582
Boulton, M 634, 920

Brewster, D 427, 428
British Association for the Advancement of Science 491
British Library MS Add.52495 589
Bromley, H 929, 930
Brougham, H P 71, 419, 422, 438-443, 458, 576, 916
Brown, A C 4, 5, 457, 464
Brown, D H 624
Brown, J 934
Brown, James 456
Brown, John 405
Bruges, P T de See De Bruges, P T
Buchanan, P D 632
Buchanan, W W 624
Buchholz, W H S See Bucholz, W H S
Bucholz, W H S 310, 324
Bucquet, J B M 215
Buess, H 555, 556
Bugge, G 500
Bugliarello, G 622
Butterfield, L H 209, 542

C See Thomson, T
Campbell, R H 636
Cardwell, D S L 600
Cavallo, T 49, 50
Cavendish, H 210, 232, 442, 443, 462
Cayley, G 570, 572, 579, 585
Chambers, R 434, 917
Christie, J R R 633
Christie, Manson & Woods 634
Christie's (auctioneers) 910
Cleghorn, G 504
Cleghorn, W 222, 568
Clerke, A M 430, 467
Clow, A 522, 543
Clow, N L 522, 543
Cochrane, A 234, 522, 615
Cochrane, T 76, 512, 901-904, 930

Cockburn, H A 483
Cockburn, H T 448, 483
Cohen, E 481, 487
Cole, W A 30, 31, 566, 634
Collison, J 219
Comenius, J A 650
Conant, J B 539a
Cook, C 918
Cookson, I 219
Cooper, J 920
Cort, H 51, 52, 234, 454
Cowen, D L 540, 560, 611
Cozens-Hardy, B 537
Craik, G L 916
Crans See Crantz, H J N von
Cranston, J A 499, 534
Crantz, H J N von 214, 306, 317
Cranz, H I N See Crantz, H J N von
Crawford, A 242, 244, 252
Crawfort, A See Crawford, A
Crell, L F F von 46, 47, 72, 74, 220a, 221a, 227, 518, 913
Crosland, M P 569, 577
Crowther, J G 578, 908
Crum Brown, A See Brown, A C
Cullen, W 14-16, 401, 402, 433, 453, 535, 569, 571, 575, 596, 609, 612, 636, 647

D, E 423a
D, K H 423b
D, L See Dobbin, L
Daiches, D 652
Dalrymple, J 234
Daniel, C F 327
Davis, A B 602, 603
Davis, T L 507
Day, S B 6
De Bruges, P T 219
De Fourcy, - See Fourcy, - de
De Luc, J A See Luc, J A de
Dean, T A 915
Deluc, J A See Luc, J A de
Dempster, G 53
Desmarets, N 215
Diergart, P 482
Dimsdale, N 589
Disciple du Docteur Black 211, 218, 229, 565

Dobbin, L 4, 5, 17, 334a, 336b, 473, 491, 523
Dobson, R 589
Doig, A 635
Dollfuss, J C 240
Doner, D B 622
Donnan, F G 486
Donovan, A L 91, 92, 596, 609, 612, 620-622, 636
Dopson, L 550
Doyle, W P 637, 643
Dreux, P F 302
Driessen, P 488
Drummond, A M 402
Duncan, A 247
Duncan, A M 626
Dundas, H 51
Dundonald, Lord See Cochrane, A
Duveen, D I 54, 532
Duveen Collection 239, 532
Dyck, D R 588

Ebstein, E 495
Edelstein, S M 554, 557, 620a
Edgar Fahs Smith Memorial Collection 461
Edinburgh Royal Blind Asylum & School 465, 470
Edinburgh Royal Infirmary 465
Edinburgh University 14, 15, 401, 402, 405, 450, 509, 587, 619, 653, 904, 906
Edinburgh University Chemical Society (1785) 508, 516, 519, 524, 530, 545, 546, 549
Edinburgh Univ Chemical Society (1875) 464
Edinburgh University Library 30
Edinburgh Univ Library MS Dc.2.76 90
Edinburgh Univ Library MS Dc.8.155 35
Edinburgh Univ Library MS Dc.8.156 34
Edinburgh Univ Library MS Gen.48D 34, 35
Edinburgh Univ Library MS Gen.874 91, 92
Eklund, J B 602, 603, 613

Emerson, R L 627
Erxleben, J C P 325, 327, 328
Ewart, J 251
Eyles, V A 583

Faraday, M 85
Farber, E 576
Faujas de Saint-Fond, B 259
Ferguson, A 79, 89, 423
Ferguson, J 38
Ferguson Collection 64
Finlayson, J 471
Fischer, J C 345
Fischer, P 220, 226
Fleck, A 535
Fleming, D 544
Fordyce, A 219
Foregger, R 561
Foster, M 475
Foster, T See Juniper, J
Fourcroy, A F de 246, 255, 257
Fourcy, - de 319, 320
Frackelton, W G 548
French, R K 593
Friday Club 483; cf 547
Fries, T M 484
Fry, J 219
Fyfe, A 267

Gahn, H 484
Garbett, S 219
G C M See M,G C
Gerstner, P A 591
Gibson, T 909a
Giobert, G A 58
Glasgow University 479, 485, 536, 592, 594, 594a, 600, 642, 910
Gowdie, J 1
Graham, A 48
Graham, H G 473a
Grand, H E Le See Le Grand, H E
Grant, J 465
Gray, J M 923, 924
Greding, J E 8
Greenaway, F 217
Gregory, J 43, 402
Greig, J Y T 44, 505
Gren, F A C 338, 341
Greyfriars Churchyard 409, 411, 456, 506
Grieve, J 649
Guerlac, H 90, 562, 563, 597, 614, 617, 638
Guyton de Morveau, L B 337

H, J G 226
Hahn, R 54
Hales, S 580
Hall, J 239, 583
Hamilton, W 445, 455
Hanson, T 6
Harcourt, W V 441, 443
Hardy, B Cozens See Cozens-Hardy, B
Harrison, W 470
Hartley, D 51
Heald, W M See Juniper, J
Heath, J 520, 911, 916
Heathcote, N H de V 511, 568
Heimann, P M 609
Henry, T 13, 217
Heron, R 246, 257
Home, F 38, 554, 557
Home, H 43, 78, 216
Hope, T C 34, 35, 637
Hopson, C R 245
Hornby, T 52a
Hume, D 42-44, 505, 615a
Hunter, W 29, 52a
Hutchins, T 48, 232
Hutton, J 52a, 256, 436, 591, 615, 620, 646, 906, 908

Industrial Museum of Scotland 451, 452
International Exhibition Glasgow 1901 476
Irvine, W, the elder 419a
Irvine, W, the younger 419a

Jacquin, N J 214, 264, 305, 307, 311, 318, 320, 339
Jameson, R 261
Jellicoe, A 51
Jewish National & University Library 620a
J G H See H, J G
Johnstone, A H 645

Jones, J 646, 652
Jones, P 652
Jorissen, W P 488
Juniper, J 405

K, W A 437
Kahlbaum, G W A 482
Kames, Lord See Home, H
Kay, J 436, 503, 520, 610, 906-908, 908a, 909
Keir, J 210, 219
Kendall, J 508, 516, 519, 524, 530, 545, 546, 549
Kennedy, D 573
Kent, A 525, 536, 910
Kirwan, R 261
Klaproth, M H 260
Klickstein, H S 26, 584
Kneass, W 914
Knickerbocker, W S 23, 496
Knight, C 916
Knight, D M 609
Knight, R & G 265
Krenger, - 321

Lambert, J H 220a
Landriani, M 225a
Lane, T 564
Langmajer, I J 330
Laplace, P S de 244, 252, 614
Lardner, D F 6
Latent 437
Lavoisier, A L 54-56, 58-60, 214, 215, 217, 221, 233, 244, 247, 250, 252, 255, 262, 263, 459-461, 468, 475, 492, 501, 526, 533, 583, 604, 614, 621, 628, 637, 640
Law, R J 911
Lawrence, C 639
L D See Dobbin, L
Le Grand, H E 607
Lee, J 34, 35
Leicester, H M 26, 584, 608
Leslie, P D 221
Lewis, W 205, 237, 247, 250, 262, 632
Lind, J 49, 50
Lindsay, R B 609a
Linnaeus, C 484

Lockhart, J G 435, 440
Lockhart, S B 905, 909
Lorenz, M 908a
Lubbock, R 235
Luc, J A de 238, 422

M, G C 45
Macbride, D 207, 214, 215, 564
Macie, J L See Smithson, J
Mackenzie, J E 509
Mackenzie, K 926
Mackenzie, R C 650
Mackie, A 615
Macquer, P J 210, 331
Maddocks, J 402
Magalhães, J J de See Magellan, J H de
Magellan, J H de 223, 564
Magie, W F 510
Marcet, A J G 79
Martin, D 904, 907, 909, 909a
Martine, G 61, 212, 249
Masson, I 493
Mavor, W F 259
M B L S 424
McElroy, D D 547
McIntyre, P 904
McKie, D 60, 76, 211, 511, 512, 517, 533, 565, 567, 568, 570, 572, 573, 579, 585, 589, 598, 629, 901-904, 930
Medico-Philosophical Society of Dublin 504
Meldrum, A N 501, 628
Mellon, P & M 618a
Meyer, J F 214, 301-303, 305-311, 316-318, 320, 333, 338-340, 344, 345, 580
Mieli, A 526
Miles, W D 209, 558
Mitchill, S L 616
Moncreiff, J 466
Monro, A, secundus 29, 648
Monro, A, tertius 606
Montesquieu, C L de S 623
Montgolfier, J 432
Morgan, J 542
Morris, R J 604

Morveau, L B Guyton de See Guyton de Morveau, L B
Moulton, F R 527
Mudie, P 79
Muirhead, J P 87, 88, 444, 447, 449
Muspratt, S 918
Musson, A E 593a

Nairne, E 221a
Naumann, W 521
Neave, E W J 513, 514, 541
Neu, J 239
Neufville, Z 332
Neumann, C 205
Neville, S 537
New York Academy of Medicine 921
Newell, L C 503
Nierenstein, M 518

Oddy, R 605
Oesper, R E 506
Oldroyd, D R 606
Osborn, J M 623
Ostwald, W 480

Pachella, R 618a
Pancaldi, G 615a
Partington, J R 517, 574, 580, 629
Paterson, J 436
Paton, H 436
Patterson, T S 538
Peacock, G 441
Pereira Pinto, G 581
Perrin, C E 235, 640, 644
Philosophical Society of Edinburgh 627
Pilatre de Rozier, F 211, 541, 565
Pinto, G P See Pereira Pinto, G
Playfair, L 450, 462
Playfair Collection 619
Poker Club 78, 483, 547
Porter, R 648
Posselwhite, J 916
Pottle, F A 582
Pratt, H T 616
Priestley, J 13, 221, 475, 564
Pringle, J 39, 564
Proks, I 650

Pyl, J T 334

R, Ch See Richet, C
Raeburn, H 910-921
Ramsay, W 50, 89, 472, 479, 480, 482, 486
Read, J 531, 539, 908
Read, W 932
Rees, A 266
Reid, T 417, 445, 455
Reilly, R 925
Reuss, A C 227
Rhees, W J 86
Ribeiro, A 623
Rice, D T 904
Richardson, B W 469, 474
Richardson, H 589
Richet, C 468
Riddell, H 489, 492
Riedel, C F 931
Robertson, - 930
Robertson, W 403
Robinson, E 593a, 598
Robison, J 36, 37, 61, 71-75, 420, 425, 427, 428, 447, 570, 572, 585, 590, 633, 912
Rogers, J 917
Roller, D 539a
Ron, M 620a
Rosmäsler, J 913
Rotheram, J 250, 262
Royal Bank of Scotland 653
Royal College of Physicians and Surgeons of Glasgow 909a
Royal College of Physicians of Edinburgh 213, 230, 248
Royal Medical Society (Edinburgh) 909
Royal Museum of Scotland 237, 605, 632, 653
See also Royal Scottish Museum
Royal Scottish Museum 610, 641
See also Royal Museum of Scotland
Royal Society of Edinburgh 466, 646
Rozier, F P de See Pilatre de Rozier, F
Rush, B 209, 542, 558
Russell, J 52a

Ryskamp, C 582

S See Scherer, A N
S, M B L See M B L S
St Andrews University 212, 531
Saint-Fond, B F de See Faujas de Saint-Fond, B
Santucci, A 615a
Savage, G 925
Schaefer, G 521
Scharf, G 919
Scheele, C W 334a, 336a, 336b, 532
Scherer, A N 264, 412, 416
Schifferes, J J 527
Schofield, R E 598, 599, 609
Schwediauer, F X See Swediauer, F X
Science and Art Department 452, 463
Science Museum (South Kensington) 494
Science Museum Library MS 371, 51
Scott, W 435, 515
Scottish National Portrait Gallery 507, 905, 909
Scottish Society of the History of Medicine 641
Sebastiani, F 618, 630, 651
Seguin, A 244, 252
Shackleton, R 623
Shannon, R 219
Sheffield, Lord 51
Sibbald, J 237
Sidney M Edelstein Collection 620a
Simpson, A D C 610, 641, 905, 906
Sinclair, J 53, 432
Skene, A 455
Skene, D 455
Skinner, A S 636
Smeaton, W A 601
Smellie, W 2, 3, 43, 465
Smellius, G See Smellie, W
Smiles, S 920
Smith, A 42-44, 256, 473a, 505
Smith, E C 497
Smith, E F 209, 928
Smith, T 402
Smithson, J 80, 86

Society for the Diffusion of Useful Knowledge 916
South Kensington Museum 462, 463
Special Loan Collection of Scientific Apparatus 462, 463
Speter, M 498, 500, 502
Spielmann, J R 304
Stanhope, Lord 51, 454
Stanier, R 928
Stanley, J T 64, 253, 254
Stansfield, D A 647
Starobinski, J 609
Stewart, D 256, 417
Stirling, W 477
Storer, J 465
Strahan, W 42
Stuart, A 522, 543
Swediauer, F X 33
Swinbank, P 594, 594a, 642

Talbot, G R 590
Tassie, J 551, 922-927
Tassie, W 923, 924
Thompson, T 419a
Thomson, D 905, 909
Thomson, J 433, 453
Thomson, T 344, 425, 426, 430, 431, 912, 915
Thornton, - 930
Thornton, R J 251
Thorpe, T E 461, 490
Tilloch, A 415
Todd, Lord 76
Todd, W B 42
Todhunter, E N 559
Trease, G E 625
Trudaine de Montigny, J C P 564
Trye, C B 647
Tytler, A F 78
Tytler, J 260a

University College MS Add.96 585
University of Minnesota Medical School 6
University of Pennsylvania 209, 461, 558

Venel, G F 255
Verbruggen, F 595

124

Vernon Sale 924
Victoria and Albert Museum 922
Vogel, R A 324
Volta, A 651

W A K See K, W A
Wall, J 225
Wall, M 225, 228
Wallace, J 571, 575
Ward, I W 478
Watson, E C 520, 551, 922
Watt, J 71, 87, 88, 219, 238, 427,
 428, 441-444, 447, 449, 454, 544,
 590, 592, 593a, 594, 594a, 598,
 600, 601, 615, 622, 920, 927
Watt Collection 63, 71, 927
Weale, J 51
Weber, J A 333, 334a, 336a, 336b
Webster, T 52, 454
Wedgwood, J 61, 925, 927
Wedgwood Museum 922
Weeks, M E 515
Weigel, C E 233, 312, 315, 322, 334
Well, J J von 312, 313, 315, 322,
 324, 330
Welleck, R 623
Wenzel, C F 329, 335
Whytt, R 203, 206, 593
Wiegleb, J C 245, 303, 309, 313,
 314, 324, 330, 340
Wightman, W P D 609
Wigney, G A 437
Williams, J 45
Williams, T I 552
Wilson, A 594
Wilson, G 446, 451, 452
Withering, W 251
Witten, L C 618a
Wood, A 448
Wordsworth, E 446

Yale University Library 618a